Praise for *Into the Magic Shop*

"*Into the Magic Shop* is pure magic! That a child from humble beginnings could become a professor of neurosurgery and the founder of a center that studies compassion and altruism at a major university, as well as an entrepreneur and philanthropist, is extraordinary enough. But it is Doty's ability to describe his journey so lyrically, and then his willingness to share his methods, that make this book a gem."

—Abraham Verghese, MD, author of *Cutting for Stone*

"Once in a generation, someone is able to articulate the compelling mystery within his or her life story in such a way that it captures the imagination of others and inspires them to align with what is deepest and best in themselves and allow it to manifest and flower. There is plenty of magic in this book, but the deepest magic of all is that Jim was openheartedly guided to start practicing that aligning when he was twelve, and trusted it enough to never lose the thread completely, even in the hardest of times. Behold what is emerging now."

—Jon Kabat-Zinn, PhD, author of *Full Catastrophe Living*

"True healing is both biological and spiritual. When you experience love and compassion, your body shifts into homeostasis and self-regulation. When you heal yourself, you heal others. The reverse is also true. Your acts of kindness and compassion are the true healing of the world. In this extraordinary book, Dr. James Doty shows you the way."

—Deepak Chopra, MD, coauthor of *Super Brain*

"Jim has written a truly extraordinary book. He shares with us the trauma of his childhood filled with pain, despair, and shame through the gifts of spirit that blazed the path for him to fulfillment, love, and wisdom. Indeed, *Into the Magic Shop* offers each of us this gift. I am in awe of Jim's skill in conveying it through the magic of this book."

—Sharon Salzberg, author of *Real Happiness*

"A compelling narrative that demonstrates the power of compassion to change not only a life but the world. Powerful and moving."

—Chip Conley, author of *Emotional Equations*

"*Into the Magic Shop* is not only a moving testimony that keeps the reader enthralled throughout the book but also a powerful exhortation to live a more compassionate and meaningful life. Beautiful and highly inspiring."

—Matthieu Ricard, author of *Altruism*

"When a top neurosurgeon opens his heart to talk about his own difficult childhood that provided recipes for loneliness, fear, rage, and shame, you know you are in for a deeply moving and insightful journey into the suffering and fragility of the human mind. Beautifully written, deeply original, this is an extraordinarily moving and exceptionally practical book into the 'magic' tricks for calming and cultivating our minds. Here are lived ways to cope and engage with the realities and struggles of life that we all just find ourselves in. The seeds of compassion are being planted; we must now cultivate them."

—Paul Gilbert, PhD, OBE, author of *The Compassionate Mind*

"*Into the Magic Shop* is a moving and inspiring story of transformation. It provides us lessons about how to live better and more compassionate lives."

—Paul Ekman, PhD, author of *Emotions Revealed*

"Dr. Doty's powerful book is a testament to how faith and compassion extend beyond religion, race, and nationality and can help an individual overcome adversity and personal limitations. It is an inspiration."

—Sri Sri Ravi Shankar, spiritual leader and founder
  of the Art of Living Foundation

"I can think of no comparable book with such a brilliantly created narrative following the remarkable arc of the author's life: From growing up as a poor, disadvantaged child into a brilliant neurosurgeon and wealthy entrepreneur, Doty's story moves deftly, from using his scalpel to save the lives of his patients to using his compassionate heart to enrich the lives of others. Profound, deeply moving, and emotionally resonating."

—Philip Zimbardo, PhD, author of *The Lucifer Effect*

"*Into the Magic Shop* will literally rewire your brain. It is a candid and personal story about a life transformed by a chance encounter in a magic shop. It is a truly optimistic and inspirational testament to the power of compassion and the ability to overcome adversity and discover your true potential."

—Glenn Beck, nationally syndicated radio host and founder of The Blaze

"This is a story of faith beyond the bounds and barriers of religion. A story of hope in the face of life's great challenges and of magic that opens the doors of potential and healing. *Into the Magic Shop* is the journey of a brain surgeon whose life is marked by success and failure and at all times laced together in a rich fabric of hope, kindness, and compassion. A book that will touch heart, soul, and mind."

—Rev. Dr. Joan Brown Campbell, director emeritus of religion
  at the Chautauqua Institute and former executive director
  of the World Council of Churches

"While we don't always get to choose what happens in life, we can choose to cultivate our compassion and wisdom as a result. In his capacity to recognize and accept that life is a tapestry of failures and successes, neurosurgeon James Doty shares the wonder and science of the relationship between head and heart—in all its pain and promise. *Into the Magic Shop* is a compelling journey about a little boy struggling with personal challenges and how the unforeseeable consequence of wandering into a magic shop transforms his life. This book will transform your life as well."

—Lisa Kristine, Lucie Award–winning humanitarian photographer

"Every so often you read a book that you can't put down until you have read the very last word. *Into the Magic Shop* is such a book. This poignant, redemptive story will take your breath away. It will make you laugh and cry, rattle your mind, break open your heart, and shake your soul as it amuses, enchants, and illuminates. Dr. James Doty, a well-known neurosurgeon, uses the twin scalpels of wisdom and compassion to operate on our consciousness. He is a surgeon of the soul—an atheist who will have you gasping 'OMG!' This book is an explosion of grace and enlightenment."

—Rabbi Irwin Kula, co-president of the National Jewish Center
  for Leadership and Learning

"Doty's memoir is as inspiring as it is riveting. In bringing a neurosurgeon's mind to matters of the heart, he sheds light not only on what matters most in life but also on how to achieve it. As you ride the ups and downs of his life with him, you can't help but share in the magic."

—David DeSteno, PhD, author of *The Truth About Trust*

"The magic of Jim's story, and the insight with which he shares his life with us all, is a great gift, and one which I encourage everyone to receive with open arms. His words, and his letters—*CDEFGHIJKL*—deserve our full attention, our full intention, and the reward is the discovery of the power of opening one's heart to each other and to the world."

—Scott Kriens, codirector of the 1440 Foundation
and chairman of Juniper Networks

"Dr. Doty's story is a captivating, archetypal tale spanning desperate heartache to a zenith of privilege and success, then landing in a billow of thoughtful, dedicated tenderness. Touched on the verge of adolescence by a strip mall fairy godmother whose magic wand was selfless, affectionate coaching on inner life and purpose, Jim transcended the common pitfalls of reflexive anxiety and diffidence—and gratuitous wealth—and with grit discovered his own love for and profound commitment to humanity."

—Emiliana R. Simon-Thomas, PhD, science director
of the Greater Good Science Center

"Stanford neurosurgeon James Doty shares with us his difficult childhood and how meeting an extraordinary woman in a magic shop at twelve changed everything. A moving and eloquent story that offers us a path to open our hearts and enlighten our minds."

—Chade-Meng Tan, author of *Search Inside Yourself*

"The truth is, Jim Doty's book was not on my reading list. Then I made the mistake of skimming the first page. I was immediately seduced by Jim's openhearted, undefended honesty in sharing a most compelling and profoundly human story. He takes us along on a mesmerizing journey, from a hardscrabble childhood to the height of human achievement. It is a story rich with inspiration, insights, and life lessons that left me wishing it would never end. Can reading someone else's life story change your own? Step *Into the Magic Shop* with Jim Doty and you'll find out."

—Neal Rogin, Emmy Award–winning writer and filmmaker,
and founding board member of Pachamama Alliance

"An intensely moving and inspiring book—a powerful example of how even when we find ourselves in the most challenging and overwhelming of circumstances, compassion can open our hearts and transform our lives."

—Sogyal Rinpoche, Buddhist monk and author of
    The Tibetan Book of Living and Dying

"Rarely has a book grabbed me so quickly and so deeply—I couldn't put this down. *Into the Magic Shop* shows us the power of living with a compassionate heart and a courageous spirit."

—Marci Shimoff, author of *Happy for No Reason, Love for No Reason,*
    and *Chicken Soup for the Woman's Soul*

"*Into the Magic Shop* is a powerful testimony of how, when we choose compassion as a defining part of who we are, magic truly begins to unfold in our lives. A most inspirational book that uplifts our spirits and opens our hearts, at a time when so much of what we hear and read seems to make us lose hope in humanity. Anyone who reads this book will be changed, for the better."

—Thupten Jinpa, PhD, author of *A Fearless Heart*

"*Into the Magic Shop*, by well-known neurosurgeon James Doty, allows us to share his remarkable story of adversity and hardship and how his life trajectory is profoundly affected by a woman in a magic shop. Her lessons change his perception of the world and his place in it, and by doing so demonstrates the power of the mind to change and the power of compassion to heal. An eloquent and powerful memoir that can change your life."

—Tim Ryan, U.S. congressman and author of *A Mindful Nation*

"In this profound and beautiful book, Dr. Doty teaches us with his life, and the lessons he imparts are some of the most important of all: that happiness cannot be without suffering, that compassion is born from understanding our own suffering and the suffering of those around us, and that only when we have compassion in our hearts can we be truly happy."

—Thich Nhat Hanh, author of *Peace Is Every Step*

Into the Magic Shop

A NEUROSURGEON'S QUEST TO
DISCOVER THE MYSTERIES OF THE BRAIN
AND THE SECRETS OF THE HEART

James R. Doty, MD

AVERY

an imprint of Penguin Random House

New York

AVERY

an imprint of Penguin Random House LLC
375 Hudson Street
New York, New York 10014

Most Avery books are available at special quantity discounts for bulk purchase
for sales promotions, premiums, fund-raising, and educational needs.
Special books or book excerpts also can be created to fit specific needs.
For details, write SpecialMarkets@penguinrandomhouse.com.

ISBN: 978-1-59463-298-3
International edition ISBN: 978-0-399-57796-3

Printed in the United States of America
1   3   5   7   9   10   8   6   4   2

BOOK DESIGN BY MEIGHAN CAVANAUGH

*Penguin is committed to publishing works of quality and integrity.
In that spirit, we are proud to offer this book to our readers;
however, the story, the experiences, and the words
are the author's alone.*

To Ruth and all those like her whose insight and wisdom are given freely.

———

To His Holiness the Dalai Lama, who continues to teach me the meaning of compassion.

To my wife, Masha,
and my children,
Jennifer, Sebastian, and Alexander,
who every day are an inspiration.

CONTENTS

———

INTRODUCTION:
BEAUTIFUL THINGS

There's a certain sound the scalp makes when it's being ripped off of a skull—like a large piece of Velcro tearing away from its source. The sound is loud and angry and just a little bit sad. In medical school they don't have a class that teaches you the sounds and smells of brain surgery. They should. The drone of the heavy drill as it bores through the skull. The bone saw that fills the operating room with the smell of summer sawdust as it carves a line connecting the burr holes made from the drill. The reluctant popping sound the skull makes as it is lifted away from the dura, the thick sac that covers the brain and serves as its last line of defense against the outside world. The scissors slowly slicing through the dura. When the brain is exposed you can see it move in rhythm with every heartbeat, and sometimes it seems that you can hear it moan in protest at its own nakedness and vulnerability—its secrets exposed for all to see under the harsh lights of the operating room.

The boy looks small in the hospital gown and is almost swallowed up by the bed as he's waiting to enter surgery.

"My nana prayed for me. And she prayed for you too."

I hear the boy's mother inhale and exhale loudly at this information, and I know she's trying to be brave for her son. For herself. Maybe even for me. I run my hand through his hair. It is brown and long and fine—still more baby than boy. He tells me he just had a birthday.

"Do you want me to explain again what's going to happen today, Champ, or are you ready?" He likes it when I call him Champ or Buddy.

"I'm going to sleep. You're going to take the Ugly Thing out of my head so it doesn't hurt anymore. Then I see my mommy and nana."

The "Ugly Thing" is a medulloblastoma, the most common malignant brain tumor in children, and is located in the posterior fossa (the base of the skull). *Medulloblastoma* isn't an easy word for an adult to pronounce, much less a four-year-old, no matter how precocious. Pediatric brain tumors really are ugly things, so I'm OK with the term. Medulloblastomas are misshapen and often grotesque invaders in the exquisite symmetry of the brain. They begin between the two lobes of the cerebellum and grow, ultimately compressing not only the cerebellum but also the brainstem, until finally blocking the pathways that allow the fluid in the brain to circulate. The brain is one of the most beautiful things I have ever seen, and to explore its mysteries and find ways to heal it is a privilege I have never taken for granted.

"You sound ready to me. I'm going to put on my superhero mask and I'll meet you in the bright room."

He smiles up at me. Surgical masks and operating rooms can be scary. Today I will call them superhero masks and bright rooms so he won't be so afraid. The mind is a funny thing, but I'm not about to explain semantics to a four-year-old. Some of the wisest patients and people I have ever met have been children. The heart of a child is wide-open. Children will tell you what scares them, what makes them happy, and what they like about you and what they don't. There is no hidden agenda, and you never have to guess how they *really* feel.

I turn to his mother and grandmother. "Someone from the team will update you as we progress. I anticipate it will be a complete resection. I don't expect any complications." This isn't just surgeon-speak to tell them what they want to hear—my plan is for a clean and efficient surgery to remove the entire tumor, while sending a small slice to the lab to see just how ugly this Ugly Thing is.

I know both Mom and Grandma are scared. I hold each of

their hands in turn, trying to reassure them and offer comfort. It's never easy. A little boy's morning headaches have become every parent's worst nightmare. Mom trusts me. Grandma trusts God. I trust my team.

Together we will all try to save this boy's life.

AFTER the anesthesiologist counts him down to sleep, I place the boy's head in a head frame attached to his skull and then position him prone. I get out the hair clippers. Although the nurse usually preps the surgery site, I prefer to shave the head myself. It is a ritual I do. And as I slowly shave the head, I think of this precious little boy and go over every detail of the surgery in my mind. I cut off the first bit of hair and hand it to the circulator to put in a small bag for the boy's mother. This is his first haircut, and while it's the last thing on his mom's mind now, I know it will matter to her later. It's a milestone you want to remember. First haircut. First tooth lost. First day of school. First time riding a bike. First brain surgery is never on this list.

I gently cut away the fine light brown strands, hoping my young patient is able to experience each of these firsts. In my mind I can see him smiling with a large gap where his front teeth should be. I see him walking into kindergarten with a backpack that's almost as big as he is slung over one shoulder. I see him riding a bike for the first time—that first thrill of freedom, pedaling feverishly with the wind in his hair. I think of my own children as I continue to clip his hair. The images and scenes of all his firsts are so clear in my mind that I can't imagine any other outcome. I don't want to see a future of hospital visits and cancer treatments and additional surgeries. As a survivor of a childhood brain tumor, he will always have to be monitored, but I refuse to see him in the future as he's been in the past. The nausea and vomiting. The falling down. The waking in the early-morning hours screaming for his mother because the Ugly Thing is compressing his brain and it hurts. There's enough heartbreak in life without adding this to the mix. I continue to gently clip his hair just enough so I can do my work. I make two dots at the base of his skull where we will make our incision, and draw a straight line.

Brain surgery is difficult, but surgery in the posterior fossa is even more so and in a small child excruciatingly difficult. This tumor is large and the work painstakingly slow and precise. Eyes looking through a microscope for hours focused on one thing. As surgeons we are trained to shut down our own bodily responses as we operate. We don't take bathroom breaks. We don't eat. We have been trained to ignore when our backs ache and our muscles cramp. I remember my first time in the operating room assisting a famous surgeon who was known not only for being brilliant but also for being a belligerent and arrogant prima donna when he operated. I was intimidated and nervous, and as I stood next to him in the operating room, sweat began pouring down my face. I was breathing heavily into my mask and my eyeglasses began steaming up. I couldn't see the instruments or even the operating field. I had worked so hard, overcome so much, and now here I was, doing surgery just like I had always imagined, but I couldn't see a thing. Then the unthinkable happened. A large drop of sweat rolled off my face and into the sterile field. He went ballistic. It should have been a highlight of my life, my first time in surgery, but instead I contaminated the surgical field and was summarily kicked out of the operating room. I have never forgotten that experience.

Today my forehead is cool and my eyesight clear. My pulse is slow and steady. Experience makes the difference, and in my operating room I am not the dictator. Or a belligerent prima donna. Every member of the team is valuable and necessary. Everyone is focused on his or her part. The anesthesiologist monitors the boy's blood pressure and oxygen, his level of consciousness, and the rhythm of his beating heart. The surgical nurse constantly monitors the instruments and supplies, making sure whatever I need is within reach. A large bag is attached to the drapes and hangs below the boy's head collecting blood and irrigation fluid. The bag is attached to a tube connected to a large suction machine and constantly measures the fluids so we know how much blood loss we have at any given moment.

The surgeon assisting me is a senior resident in training

and new to the team, but he is just as focused on the blood vessels, and brain tissue, and minutiae of removing this tumor as I am. We can't think about our plans for the next day, or hospital politics, or our children, or our relationship trouble at home. It's a form of hypervigilance, a single-pointed concentration almost like meditation. We train the mind, and the mind trains the body. There's an amazing rhythm and flow when you have a good team—everyone is in sync. Our minds and bodies work together as one coordinated intelligence.

I am removing the last piece of the tumor, which is attached to one of the major draining veins deep in the brain. The posterior fossa venous system is incredibly complex, and my assistant is suctioning fluids as I carefully resect the final remnant of the tumor. He lets his attention wander for a second, and in that second his suction tears the vein, and for the briefest moment everything stops.

Then all hell breaks loose.

The blood from the ripped vein fills the resection cavity, and blood begins to pour out of the wound of this beautiful little boy's head. The anesthesiologist starts yelling that the child's blood pressure is rapidly dropping and he can't keep up with the blood loss. I need to clamp the vein and stop the bleeding, but it has retracted into a pool of blood, and I can't see it. My suction alone can't control the bleeding and my assistant's hand is shaking too much to be of any help.

"He's in full arrest!" the anesthesiologist screams. He has to scramble under the table because this little boy's head is locked in a head frame, prone, with the back of his head opened up. The anesthesiologist starts compressing the boy's chest while holding his other hand on his back, trying desperately to get his heart to start pumping. Fluids are being poured into the large IV lines. The heart's first and most important job is to pump blood, and this magical pump that makes everything in the body possible has stopped. This four-year-old boy is bleeding to death on the table in front of me. As the anesthesiologist pumps on his chest, the wound continues to fill with blood. We have to stop the bleeding or he will die. The brain consumes 15 percent of the outflow of the heart and can survive only minutes after the heart stops. It needs blood and, more important, the oxygen that is in the blood. We are running out of time before the brain dies—they need each other— the brain and the heart.

I am frantically trying to clamp the vein, but there's no way to see the vessel through all the blood. Although his head is fixed into position, the chest compressions are moving it ever so slightly. The team knows and I know that we are running out of time. The anesthesiologist looks up at me and I see the fear in his eyes. . . . We might lose this child. Cardiopulmonary resuscitation (CPR) is like trying to clutch-start a car in second gear—it's not very reliable, especially as we are continuing to lose blood. I am working blind, so I open my heart to a possibility beyond reason, beyond skill, and I begin to do what I was taught decades ago, not in residency, not in medical school, but in the back room of a small magic shop in the California desert.

I calm my mind.

I relax my body.

I visualize the retracted vessel. I see it in my mind's eye, folded into this young boy's neurovascular highway. I reach in blindly but knowing that there is more to this life than we can possibly see, and that each of us is capable of doing amazing things far beyond what we think is possible. We control our own fates, and I don't accept that this four-year-old is destined to die today on the operating table.

I reach down into the pool of blood with the open clip, close it, and slowly pull my hand away.

The bleeding stops, and then, as if far away, I hear the slow blip of the heart monitor. It's faint at first. Uneven. But soon it gets stronger and steadier, as all hearts do when they begin to come to life.

I feel my own heartbeat begin to match the rhythm on the monitor.

Later, in post-op, I will give his mother the remnants from his first haircut, and my little buddy will come out of the anesthetic a survivor. He will be completely normal. In forty-eight hours, he will be talking and even laughing, and I will be able to tell him that the Ugly Thing is gone.

PART ONE

Into the Magic Shop

ONE

Real Magic

Lancaster, California, 1968

The day I noticed my thumb was missing began like any other day the summer before I started eighth grade. I spent my days riding my bicycle around town, even though sometimes it was so hot the metal on my handlebars felt like a stove top. I could always taste the dust in my mouth—gritty and weedy like the rabbitbrush and cacti that battled the desert sun and heat to survive. My family had little money, and I was often hungry. I didn't like being hungry. I didn't like being poor.

Lancaster's greatest claim to fame was Chuck Yeager breaking the sound barrier at nearby Edwards Air Force Base some twenty years earlier. All day long planes would fly overhead, training pilots and testing aircraft. I wondered what it would be like to be Chuck Yeager flying the Bell X-1 at Mach 1, accomplishing what no human had ever done before. How small and desolate Lancaster must have looked to him from forty-five thousand feet up going faster than anyone ever thought possible. It seemed small and desolate to me, and my feet were only a foot above the ground as I pedaled around on my bike.

I had noticed my thumb missing that morning. I kept a wooden box under my bed that had all my most prized possessions. A small notebook that held my doodles, some secret poetry, and random crazy facts I had learned—like twenty banks are robbed every day in the world, snails can sleep for three years, and it's illegal to give a monkey a cigarette in Indiana. The box also held a worn copy of Dale Carnegie's *How to Win Friends and Influence People*, dog-eared on the pages that listed the six ways to get people to like you. I could recite the six things from memory.

1. Become genuinely interested in other people.
2. Smile.
3. Remember that a person's name is, to that person, the sweetest and most important sound in any language.
4. Be a good listener. Encourage others to talk about themselves.

5. Talk in terms of the other person's interest.
6. Make the other person feel important—and do it sincerely.

I tried to do all of these things when I talked to anyone, but I always smiled with my mouth closed because when I was younger I had fallen and hit my upper lip on our coffee table, knocking out my front baby tooth. Because of that fall my front tooth grew in crooked and was discolored a dark brown. My parents didn't have the money to get it fixed. I was embarrassed to smile and show my discolored crooked tooth, so I tried to keep my mouth closed at all times.

Besides the book, my wooden box also had all my magic tricks—a pack of marked cards, some gimmicked coins that I could change from nickels into dimes, and my most prized possession: a plastic thumb tip that could hide a silk scarf or a cigarette. That book and my magic tricks were very important to me—gifts from my father. I had spent hours and hours practicing with that thumb tip. Learning how to hold my hands so it wouldn't be obvious and how to smoothly stuff the scarf or a cigarette inside it so that it would appear to magically disappear. I was able to fool my friends and our neighbors in the apartment complex. But today the thumb was missing. Gone. Vanished. And I wasn't too happy about it.

My brother, as usual, wasn't home, but I figured maybe he had taken it or at least might know where it was. I didn't know where he went every day, but I decided to get on my bike and go looking for him. That thumb tip was my most prized possession. Without it I was nothing. I needed my thumb back.

I WAS RIDING through a lone strip mall on Avenue I—an area not on my usual bike circuit because apart from the strip mall there was nothing but empty fields and weeds and chain-link fences for a mile on either side. I looked at a group of older boys in front of the small market but didn't see my brother. I felt relieved because usually if I found my brother in a group of kids it meant he was getting picked on and I would be getting into a fight to defend him. He was a year and a half older than me, but he was smaller, and bullies like to pick on those who can't defend themselves. Next to the market was an optometrist's office and next to that was a store I had never seen before—Cactus Rabbit Magic Shop. I stopped on the sidewalk in front of the strip mall and stared across the parking lot. The entire storefront was five vertical glass panes with a glass door to the left. The sun glinted off the dirt-streaked glass, so I couldn't see if anyone was inside, but I walked my bike to the front door hoping it was open. I wondered if they sold plastic thumbs and for how much. I didn't have any money, but it couldn't hurt to look. I leaned my bike against a post in front of the store with a quick glance down at the group of boys in front of the market. They didn't seem to have noticed me, or my bike, so I left it there and pushed against the front door. It didn't budge at first, but then, as if by a magician's wave of his wand, it gave way and opened smoothly. A little bell rang above my head as I walked in.

The first thing I saw was a long glass counter full of packs of cards and wands and plastic cups and gold coins. Against the walls were heavy black cases that I knew were used for stage magic, and large bookcases filled with books about magic and illusion. There was even a mini guillotine in the corner and two green boxes that you could use to saw a person in half. An older woman with wavy brown hair was reading a paperback book, her glasses perched at the end of her nose. She smiled, still looking down at her book, and then she took her glasses off and lifted her head and stared straight into my eyes the way no adult ever had before.

"I'm Ruth," she said. "What's your name?"

Her smile was so big and her eyes so brown and kind that I couldn't help but smile back at her, forgetting completely about my crooked tooth.

"I'm Jim," I said. Until that moment I was called Bob. My middle name is Robert. I can't remember why I was called Bob. But for whatever reason, when she asked I replied, "Jim." And this was the name I would go by for the rest of my life.

"Well, Jim. I'm so glad you walked in."

I didn't know what to say back and she just continued to stare into my eyes. Finally, she gave a sigh, but it was more of a happy sigh than a sad kind of sigh.

"What can I do to help you?"

My mind went blank for a second. I couldn't remember why I had come into this store and I felt that same feeling you get when you lean too far back in a chair and suddenly catch yourself right before it tips all the way over. She waited patiently, still smiling, until I found the words to answer.

"My thumb," I said.

"Your thumb?"

"I lost my plastic thumb tip. Do you have any?"

She looked at me and kind of shrugged her shoulders as if she had no idea what I was talking about.

"For my magic. It's a magic trick. You know, a TT, a plastic thumb tip."

"I'll tell you a little secret," she said. "I don't know anything about magic tricks." I looked around at the endless display of gadgets and tricks of every kind and then looked back at her, no doubt surprised. "My son owns the store, but he's not here at the moment. I'm just sitting here reading, waiting for him to return from an errand. I know absolutely nothing about magic or thumb tip tricks, I'm sorry to say."

"That's OK. I'm just looking anyway."

"Of course. Feel free to look around. And then be sure to tell me if you find what you're looking for." She laughed, and while I wasn't sure why she was laughing, it was a nice laugh that made me feel happy inside for no real reason.

I wandered around the store looking at the endless rows of magic cards and props and books. There was even a display case full of plastic thumbs. I could feel her eyes on me as I browsed, and while I knew she was staring, it wasn't in the same way the guy who owns the market next to our apartment would stare at you when you were in his store. I'm pretty sure he thought I was going to steal something, and every time I went in there I could feel his suspicious eyes tracking my every step.

"Do you live in Lancaster?" Ruth asked.

"Yes," I said, "but on the other side of town. I was just riding around looking for my brother and I saw your store and decided to come in."

"Do you like magic?"

"I love it," I said.

"What do you love about it?"

I wanted to just say that I thought it was cool and fun, but something else altogether came out of my mouth. "I like being able to practice at something and get really good at it. I like that I am in control. Whether the trick works or it doesn't work is only up to me. It doesn't matter what anyone else says or does or thinks."

She was quiet for a few moments, and I immediately felt embarrassed that I had said all that.

"I understand what you mean," she said. "Tell me about the thumb trick."

"Well, you put the thumb tip on your thumb and the audience thinks it's your real thumb. You have to kind of hide it a bit, because it looks really fake if you take a good look at it. It's hollow inside and you can move it from your thumb into the palm of your other hand like this." I made a classic magic gesture—grabbing one hand with my other hand and sliding my fingers across each other. "You move the thumb tip secretly into your other hand and you can stuff a small silk scarf or a cigarette inside it and then make the moves again and put the thumb back on your finger. But now deep inside is whatever you're concealing. It looks like you made something magically disappear or you can use it the opposite way and make it look like you've made something magically appear out of thin air."

"I see," said Ruth. "How long have you practiced these tricks?"

"A few months. Every day I practice, sometimes for a few minutes, other times an hour. But every day. It was really hard at first, even with the instruction book. But then it just gets easier and easier. Anyone could do it."

"It sounds like a good trick, and that's great that you practiced, but do you know why it works?"

"What do you mean?" I asked.

"Why do you think this trick works on people? You said the thumb looks really fake, so why does it trick people?"

She looked suddenly very serious, and like she really wanted me to teach her something. I wasn't used to anyone, especially not an adult, asking me to explain or teach them anything. I thought about it for a minute.

"I guess it works because the magician is so good he fools people. They don't see his sleight of hand. You have to keep people distracted when you do magic."

She laughed at this. "Distracted. That's just perfect. You're very wise. Would you like to hear why I think the magic trick works?" She waited for me to answer, and again it felt strange to have an adult ask my permission to tell me something.

"Sure."

"I think the magic trick works because people see only what they think is there rather than what's actually there. This thumb tip trick works because the mind is a funny thing. It sees what it expects to see. It expects to see a real thumb, so that's what it sees. The brain, as busy as it can be, is actually very lazy. And yes, it also works because people are, as you said, so easily distracted. But they're not distracted by hand gestures. Most people who are watching a magic show aren't really there watching the magic show. They are regretting something they did yesterday or worrying about something that might happen tomorrow, so they're not really at the magic show to begin with, so how could they see the plastic thumb at all?"

I didn't really understand what she was saying, but I nodded my head. I would have to think about this later. Replay her words in my head and puzzle it out.

"Don't get me wrong. I believe in magic. But not the kind that requires gimmicks and tricks and sleights of hand. Do you know the kind of magic I'm talking about?"

"No. But it sounds cool," I said. I wanted her to keep talking. I liked that we were having a real conversation. I felt important.

"Do you ever do any tricks with fire?"

"Well, you can do the thumb trick with a lit cigarette also, but I haven't done it that way. You have to use fire to light the cigarette."

"Well, imagine that there was a little flickering light and you had the power to make it turn into a giant flame, like a fireball."

"That sounds really cool. How do you do it?"

"That's the magic. You can turn this little tiny light into a huge fireball with only one thing—your mind."

I didn't know what she meant, but I liked the idea a lot. I liked magicians who could hypnotize people. Bend spoons with their mind. Levitate.

Ruth clapped her hands together.

"I like you, Jim. I like you a lot."

"Thanks." It felt good to hear her say that.

"I'm going to be in this town for only six weeks, but if you agree to come see me every day for the next six weeks, I will teach you some magic. The kind of magic that you can't buy in a store and that will help you make anything you want actually appear. For real. No tricks. No plastic thumbs. No sleights of hand. How does that sound?"

"Why would you do that?" I asked.

"Because I know how to turn a flicker into a flame. Someone taught me and now I think it's time that I teach you. I can see the special in you, and if you come here every day, without missing a single day, you're going to see it too. I promise. It's going to take a lot of work and you're going to have to practice the tricks I teach you even more than you did your thumb trick. But I promise you, what I'm going to teach you will change your life."

I didn't know what to say to that. No one had ever called me special. And I knew if Ruth knew the truth about my family and who I was, she wouldn't think I was special at all. I didn't know if I believed she could teach me to make things appear out of nowhere, but I wanted to have more conversations with her like we had today. Being around her made me feel good inside. Happier. Almost like I was loved, which I knew was a weird thing to feel from a complete stranger. She looked like she could be anyone's grandmother, except for her eyes. Her eyes promised mystery and secrets and adventure. There was no other adventure waiting for me this summer, and here was this woman offering to teach me something that could change my life. How weird. Whether she could or not, I didn't know, but what I did know is that I had absolutely nothing to lose. I felt hope, something I hadn't felt a lot of before.

"What do you say, Jim, are you ready to learn some real magic?"

And with that simple question the entire trajectory of my life and whatever fate had previously had in store for me shifted.

TWO

A Body at Rest

Since the start of civilization, the source of human intelligence and consciousness has been a mystery. In the seventeenth century B.C.E., the Egyptians believed that intelligence resided in the heart. Upon death, it was the heart that was revered and retained with the other internal organs. The brain had so little value to the ancient Egyptians that it was routinely removed with a hook through the nasal cavity before mummification, and then discarded. In the fourth century B.C.E., Aristotle believed that the brain functioned primarily as a cooling mechanism for the blood, and this was why humans (with their larger brains) were more rational than the "hot-blooded" beasts. It took five thousand years for this view of the brain's insignificance to be reversed. The brain's centrality to our identity began to be understood only because individuals who had sustained head trauma through accident or war injuries demonstrated impairment of thought or function. While much was learned about brain anatomy and function, our understanding remained very limited. In fact, through most of the twentieth century, it was believed that the brain was fixed, immutable, and static. Today we know that the brain has great plasticity and can change, adapt, and transform. It is molded by experience, repetition, and intention. It is only because of the extraordinary technological advances over the last few decades that we can see the brain's ability to transform on a cellular, genetic, and even molecular level. Extraordinarily, as I learned, each of us has the ability to change the very circuitry of our brain.

My first experience of neuroplasticity happened with Ruth in the back room of that magic shop in a strip mall. I didn't know this at age twelve, but during those six weeks she literally rewired my brain. She did what, at that time, many would have said was impossible.

I DIDN'T TELL anyone about my plans to go to the magic shop every day, but then again, no one really asked. Summer in Lancaster was like being in some hot, windswept, seemingly endless purgatory—there was always a restless feeling that I should be doing something, but there was nothing really to do. The apartment complex I lived in was surrounded by little more than packed earth and tumbleweeds. Occasionally this landscape was interspersed with an abandoned car or a derelict piece of machinery. A thing no longer wanted or needed—thrown away in a place no one would notice.

Children, and adults as well, perform best when there is consistency and dependability. The brain craves them both. In my house there was neither. No set time to eat, no alarm to remind you to wake up for school, and no bedtime. If my mother's depression lessened enough for her to leave the bed, a meal might be made. That is if there was food in the house. If not, I would go to sleep hungry or go visit a friend and hope he would ask me to stay for dinner. I thought I was lucky because, unlike most of my friends, I never had to be home at a particular time. I didn't want to get home until late because I knew if I got home earlier there would often be a fight in progress or some other event that made me wish I were somewhere else, someone else. Sometimes the thing you want most is just someone to tell you, tell you anything. Because that means you're important. And sometimes it's not that you're not important, it's just you're not seen because the pain of those around you makes you invisible. I pretended I was lucky because I didn't have anyone to bother me—to tell me to do my homework, wake me up for school, or tell me what to wear. But I was only pretending. Teenagers crave freedom, but only if they're standing on a base that is stable and secure.

. . .

RUTH HAD ASKED ME to come to the shop at 10 A.M., and I woke up early that first day feeling as if it were my birthday and Christmas morning both rolled into one. I had had a hard time going to sleep. I had no idea what she was going to teach me, and I didn't really care. I just wanted to talk to her some more, and it felt good to have somewhere to go. I felt important.

I COULD SEE RUTH through the window of the magic shop that first day as I rode up on my orange Schwinn Sting-Ray with its white banana seat. I remember that bike so well because it was the most valuable thing I had, and I had bought it with my own money. Money made from mowing lots of lawns in the heat of those long summer days. As I pulled up I saw that she wore a big blue headband that held her shoulder-length brown hair away from her face, and her glasses dangled on a chain at her neck. Her dress looked almost like the big smocks we had to put on over our clothes during art class at school. It was exactly the same color as the Lancaster sky in the morning—the lightest of blues with horizontal wisps of white. Every morning when I woke up, the first thing I did was look out my window. For some reason seeing that blue sky always made me feel hopeful.

Ruth gave me a big smile, and I smiled back, but I could feel my heart hammering in my chest. I knew it was partly from riding there so quickly and partly because I wasn't sure what was going to happen. And I didn't know why it was happening. It had sounded like a good idea the day before, and this morning it had seemed better than another day riding through endless fields of tumbleweeds on my Sting-Ray, going nowhere but always hoping to end up somewhere. Yet at that moment I wasn't so sure.

What was I walking into? What if I wasn't smart enough to learn whatever magic she was going to teach me? What if she found out the truth about my family? What if she was really some crazy lady who was going to kidnap me and take me out into the middle of the desert and do black magic with my dead body? I had seen a movie called *Voodoo Woman* a while back and suddenly I wondered if Ruth was a mad sorceress who was going to turn me into a monster she could control with her mind and then take over the world.

My arms went weak. I had the door halfway open, but suddenly it felt heavy. It resisted me. I looked at my bike lying on its side and at the nearly empty parking lot. What was I doing? Why had I agreed to this? I could get on my bike and ride away and never come back.

Ruth smiled and called out my name. "Jim, it's so good to see you. For a minute I wasn't sure you were going to come." She nodded her head in a grandmotherly way and waved at me to come in. I felt warm inside. She didn't seem like a mad sorceress bent on my destruction.

I pushed open the door the rest of the way. It now easily swung open.

"You were riding that bike down the street like you were being chased," she said as I stepped inside. I often felt like I was being chased, but I did not know who was chasing me.

Suddenly my face flushed with shame. Maybe she saw my fear or my doubts. Maybe she had X-ray vision. I looked down at my old tennis shoes. There was a small hole at the top of my right shoe. I was embarrassed. I curled my toes so she wouldn't see them.

"This is my son, Neil. He's the magician." If she had noticed the hole in my shoe, she was hiding it.

Neil didn't look like a magician, really. He had big black glasses and the same shade of brown hair as his mother. He seemed normal enough. No magic hat, no cape, no mustache.

"So, I hear you like magic." Neil's voice was deep and slow. He had what looked like fifty decks of cards stacked on the glass countertop.

"Yeah, it's pretty cool."

"Do you know any card tricks?" Neil started shuffling a deck in his hands. The cards seemed to fly from his right hand to his left hand, back and forth, back and forth, flying through space. I wanted to learn how to do that. He stopped and fanned the deck out in front of me.

"Pick a card."

I looked at the cards. One card was sticking out slightly, so I figured that was the obvious choice and instead picked a card way off to the right.

"Now don't show me what it is, but hold it in front of you and take a look."

I glanced down, keeping the card close to my chest, just in case there were any mirrors behind me. It was the queen of spades.

"Now put it facedown anywhere in the deck, and now I want you to shuffle the cards. Mix them up any way you want. Here you go."

Neil handed me the entire deck, and I tried to shuffle them—not like he had, but I managed to get my hands around the cards and did a pretty good job of shuffling them.

"Shuffle them again."

I did it again and was a little better this time. The cards stacked a little more crisply and orderly.

"Now a third time."

This time I remembered to push my knuckles into the cards to make them bow, and when they came together they were like two gears that spun together.

"Very good." I handed the deck back to him. He started turning the cards over one by one, faceup. Every once in a while he would hold up a card and say, "This is not your card." Finally he turned over the queen of spades.

"This is the one. This is your card." He waved the card around with a flourish and set it faceup in front of me on the counter.

"That's cool," I said with a smile, wondering how he knew it was my card. I picked it up and turned it over. I checked all four sides of the card to see if it was bent at all. Nothing.

"Do you know who this is? Who the queen of spades represents?"

I tried to remember the name of a queen I had heard about in history class. I could only remember one. "Queen Elizabeth?"

Neil smiled at me. "Now if this were an English deck of cards you would be correct. But this happens to be a French deck of cards, and in the French deck, each queen represents a different woman in history or mythology. The queen of hearts and the queen of diamonds in a French deck represent Judith and Rachel, both powerful women from the Bible. The queen of clubs is known as Argine, who isn't anyone I've ever heard of, but her name is an anagram of *Regina*, which is Latin for 'queen.' The queen of spades, your card, is the Greek goddess Athena. She is the goddess of wisdom and the companion of all heroes. If you go on a heroic quest, you definitely want Athena on your side."

"So how did you know that was my card?"

"Now you know a magician never reveals his tricks, but given that you are here to learn, I guess I can let you in on the secret." Neil turned the card over. "This deck happens to be a marked deck. It looks like a regular Bicycle deck, but if you look closely at what looks like a flower here at the bottom, you can see that there are eight petals around the center. Each petal represents a card from two to nine, and the center of the flower represents the ten. Over to the side, these four swirls represent the suits." He pointed to another design to the side of the flower. "When magicians mark decks, we shade in either a petal or the center and a petal to represent the jack, queen, and king. If nothing's shaded it's an ace. And then we mark up here to show the suit. So if you look at your card, you can see the code. The center is shaded plus the number three petal, so that is a queen. And over here you can see that the spade is shaded."

I studied the card. The shading was subtle, and if I didn't know what I was looking for I never would have noticed.

"It takes some study, but once you've memorized it, you can read them quickly."

I looked at all the other decks spread out on the counter. "Are all of these decks marked?"

"No. These are all different types of trick decks. Stripper decks. Svengali decks. Gaffed decks. Forcing decks. I even have a brainwave deck. I make them all. Cards are my specialty."

I had heard of gaffed decks, with trick cards like the thirteen of diamonds or a dead king of spades or a joker who is holding the exact card someone in the audience picks, but that was it. All the other names sounded so mysterious. Stripper decks and brainwave decks? I had no idea what these could be, but I didn't want to admit my ignorance to Neil.

"You know in World War II there were special decks of cards that were made and sent to prisoners of war in Germany? Each card could be peeled apart and hidden inside was a section of a map that, if you pieced it all together, showed a secret escape route for the prisoners. Now that was an amazing magic trick."

Neil put the queen of spades back in the marked deck and handed it to me. "You can have this. It's a gift."

I took the deck from him. No one ever just gave me things for free. "Thanks," I said. "Thanks a lot." I vowed to myself to memorize every marked card.

"So, my mother tells me she's going to teach you some really cool magic."

I smiled, not sure what to say.

"Her magic is way beyond anything we have out here." He waved his hand around the store. "With her magic you can learn to get anything you want. It's kind of like a genie in a bottle, but she's going to introduce you to the genie in your head. Just be careful what you wish for."

"Three wishes?" I asked.

"As many wishes as you want. It's going to take a lot of practice, though. It's much harder than learning card tricks, but it might not look like much. I had to practice for a really long time. Just remember to pay close attention to everything she says. There aren't any shortcuts. You have to follow every step exactly as she tells you."

I nodded at Neil and put the marked deck of cards in my pocket.

"She's going to take you in the back now. We have a small office back there. Remember, do whatever she tells you." He looked over and smiled at her.

Ruth patted her son on the forearm and then looked at me. "C'mon, Jim. Let's get started."

She walked toward a door in the back wall of the store and I followed, having no idea what I was really doing.

THE BACK OFFICE was dim and smelled a bit musty. There were no windows and only an old brown desk and two metal chairs. The carpet was brown shag, matted down in the middle of the room and sticking up like short brown grass around the walls. There were no magic tricks in here. No wands or plastic cups or cards or hats.

"Sit down, Jim."

Ruth sat in one of the metal chairs and I sat down in the other one. We were face-to-face and our knees were almost touching. My right leg was jittering up and down, which it did every time I was nervous. My back was to the door but I knew where it was in case I needed to run. I mentally calculated how long it might take me to get out of there and get to my bike.

"I'm glad you came back today." Ruth smiled at me, and I felt a little less jittery. "How do you feel?"

"OK."

"What are you feeling right now?"

"I don't know."

"Are you nervous?"

"No," I lied.

Ruth put her hand on my right knee and pressed down. My knee immediately stopped moving. I braced myself, ready to run if this got any weirder. She took her hand off my knee.

"You were shaking your leg like you were nervous."

"I guess I'm just wondering what you're going to teach me."

"The magic I'm going to teach you is not something you can buy in a store. This magic has been around for hundreds, maybe thousands, of years and you can learn it only if somebody teaches it to you."

I nodded my head.

"But you have to give me something first."

I was pretty sure I would have given Ruth anything to learn her secrets, but other than my bike I didn't have much.

"What do you want?"

"You need to promise me that you will teach someone else what I'm teaching you this summer. And you have to get that person to promise they will teach someone. And so on. Can you do that?"

I had absolutely no idea who I would teach, and at that point I didn't even know whether I could teach it to someone else. But Ruth was just staring at me, waiting for an answer, and I knew there was only one right answer.

"I promise."

I thought about crossing my fingers behind my back just in case I couldn't find anyone to teach, but instead raised three fingers up in the air like I had seen the Boy Scouts do. I figured that made it official.

"Close your eyes. I want you to imagine you are a leaf blowing in the wind."

I opened my eyes and grimaced. I was really tall for my age but only about 120 pounds. I was more like a twig stuck in the ground than a leaf blowing in the wind.

"Close your eyes," she said kindly, and nodded.

I closed my eyes again and tried to imagine a leaf blowing in the wind. Maybe she was going to hypnotize me into thinking I was a leaf. I had seen a stage hypnotist before, and he had made people from the audience think they were different farm animals. Then he had them all start fighting with each other. I started laughing and opened my eyes.

Ruth sat upright in the chair across from me, her hands resting palms-down on her thighs. She sighed a little.

"Jim, the first trick is to learn how to relax every muscle in your body. It's not as easy as it sounds."

I wasn't sure I ever felt relaxed. It seemed like I was always ready to either run or fight. I opened my eyes again, and Ruth tilted her head to the right and looked straight into my eyes.

"I'm not going to hurt you. I'm going to help you. Can you trust me?"

I thought about what she was asking. I didn't know if I trusted anyone in my life, certainly not adults. But no one had ever asked me to trust them before, and I liked the way it felt. I wanted to trust Ruth. I wanted to learn what she had to teach me, but the whole situation felt so weird.

"Why?" I asked. "Why are you going to help me?"

"Because the second we met I knew you had potential. I see it. And I want to teach you to see it too."

I didn't know what potential was or how she knew I had it. I also didn't know then that she probably would have seen potential in anyone who had strolled into the magic shop on that summer day in 1968.

"OK," I said. "I trust you."

"Good, then. That's where we will begin. Focus on your body right now. How does it feel?"

"I don't know."

"Think about riding your bike. What does it feel like in your body when you ride your bike really fast?"

"It feels good, I guess."

"What is your heart doing right now?"

"Beating," I said, and smiled.

"Slow or fast?"

"Fast."

"Good. How do your hands feel?"

I looked down and noticed my hands were gripping on to the edge of the chair. I relaxed them.

"They're relaxed."

"OK. What about your breathing? Is it deep or shallow?" Ruth took a deep breath in and out. "Like this or like this?" She started breathing fast, like a panting dog.

"I guess it's somewhere in between."

"Are you nervous?"

"No," I lied.

"Your leg is shaking again."

"Maybe a little."

"The body is full of signs about what's going on inside us. It's really amazing. Someone can ask you how you are feeling and you might say, 'I don't know,' because maybe you don't know or maybe you don't want to say, but your body always knows how you are feeling. When you are afraid. When you are happy. When you are excited. When you are nervous. When you are angry. When you are jealous. When you are sad. Your mind might think you do not know, but if you ask your body, it will tell you. It has a mind of its own, in a way. It reacts. It responds. Sometimes it reacts the right way in a situation, sometimes the wrong way. Do you understand?" I suddenly thought about how that was true. When I would come home, I could tell immediately what mood my mother was in as soon as I walked in the door. She didn't have to say a word. I could feel it in the pit of my stomach.

I shrugged. I was trying to follow what she was saying.

"Do you ever get really sad or really angry?"

"Sometimes." I was angry a lot, but I didn't want to say so.

"I want you to tell me about an instance when you were angry or afraid—and then we're going to talk about how it feels in your body when you tell me about it."

My mind started racing. I didn't know what to tell her. Should I tell her about the time I went to Catholic school and the nun slapped me and, without thinking, I slapped her back? Or maybe about Thursday night when my dad came home drunk again? Or I could tell her about what that doctor said when I took my mother to the hospital and how it made me want to hit him or crawl into a hole or both.

"Jim, your thoughts are so loud I can hear them, but I can't make them out. Tell me what you're thinking right at this very moment."

"I'm thinking about all the things I don't want to tell you."

She smiled. "It's OK. There's nothing you can say that would be wrong. We're talking about what you felt. Feelings are not right or wrong. They are just feelings."

I didn't really believe what she was saying. I felt an enormous amount of shame about my feelings, my anger, my sadness, all the ways my emotions seemed to take me over. I wanted to run away.

"Your leg is going up and down a mile a minute right now," she said. "I'm going to count to three, and I just want you to start telling me a story. You're not going to think about what you're going to say, OK? I'm going to count to three. Ready?"

I was still frantically trying to clear away all of my erupting thoughts and feelings to find something that was not so embarrassing. I didn't want to scare her away.

"One . . ."

What if she were Catholic and would be horrified to hear I slapped a nun and was kicked out of school and sent away to live with my older sister and her husband, where I also got in a fight and got kicked out of that school? What if she didn't want me to come back because I was too violent?

"Two . . ."

What if I told her about how mad I was at my dad for getting drunk and wrecking our car, and now we had to drive around with the entire front end dented in and the bumper held on with rope and it was like a big sign that said *look how poor we are, we can't even afford to get our car fixed*? What if she thinks I am a bad son?

"Three . . . Go!"

"My dad drinks. Not every day, but a lot. He will go

drinking and disappear sometimes for weeks, and we will be left without any money except for the public assistance checks we get, and they don't cover too much. When he's not drinking, we are all tiptoeing around the house trying not to set him off. When he does drink at home he yells and swears and breaks things, and my mom starts to cry. My brother disappears when this happens, and I hide out in my room, but I am always listening in case things get really bad and I need to do something. I worry about my mom. She's sick a lot and in bed almost all the time, and she always seems to get worse after he drinks, and they fight. She yells at him when he's home, and then when he leaves she goes silent. She doesn't get out of bed or eat or do anything. I don't know what I'm supposed to do."

"Go on, Jim." She was really listening. She really seemed to want to hear what I was saying. She didn't look like she was shocked. She was smiling that chocolate chip cookie smile of understanding. Like she knew what I was talking about or at least she didn't think my family was dirt because we were so poor. "Go on," she said, encouraging me.

"One time I came home from school and everything was quiet. A weird kind of quiet. I went in my mom's room, and she was in bed. She had taken a bunch of pills. They're pills to calm her down, but she took too many. I had to run to the apartment next door and ask the lady there to drive us to the hospital. She's had to go to the hospital before like this. My mom, I mean—she's done this kind of thing before. At the hospital my mom is in the bed, and I'm sitting next to her, and I can hear them talking on the other side of the curtain. One guy is so mad that they have to do all this paperwork for my mom, and he says she's been here before, and he's tired of wasting time on these types of people. The woman laughs and says something about 'maybe this will be the last time.' I can't really make it out and then they both laugh, and I am so mad I just want to tear down the curtain and scream at them. People in a hospital shouldn't be like that. And I'm mad at my mom, because I don't understand why she has to do this. It's not fair and it's embarrassing, and I'm mad at my dad for making her so angry and so sad. I'm angry at both of them and everyone at the hospital and I get really, really, mad sometimes."

I'm not sure what to do now that I've stopped talking. Ruth is sitting across from me in her chair, and I just stare down at the stupid hole in my stupid tennis shoe.

"Jim." Ruth says my name softly. "How do you feel in your body right at this very moment?"

I shrug my shoulders. I wonder what she thinks of me now that she knows about my family.

"How does your stomach feel?"

"Kind of sick."

"How does your chest feel?"

"Tight. It hurts a little."

"What about your head?"

"My head is pounding."

"How about your eyes?"

I don't know why, but the minute she asked that question I felt like all I wanted to do was close my eyes and cry. I wasn't going to cry. I didn't want to cry, but I couldn't help it. A tear rolled down my check.

"My eyes are stinging a bit, I guess."

"Thank you for telling me about your parents, Jim. Sometimes we need to stop thinking about what we should say and just say what it is we need to."

"Easy for you to say."

Ruth and I both laughed and in that second I felt a little better.

"My chest doesn't feel as tight."

"Good. That's good. I'm going to teach you how to relax every muscle in your body and I want you to practice this every day for an hour. Everything we practice here every morning I want you to also practice at home at night, sort of like homework. Now, relaxing may sound easy, but it's actually very, very hard to do, and it takes a lot of practice."

I still wasn't sure I could remember a time when I felt relaxed. I've felt tired plenty of times, but I don't know if I've ever felt relaxed. I wasn't even sure what it meant.

Ruth told me to sit in the chair in a comfortable position and close my eyes. She asked me again to imagine I was a leaf blowing in the wind. It felt kind of cool to soar around the streets in my head. I felt a little lighter in the chair.

"Don't slouch. You want to stay awake and you want to keep your muscles engaged even though you're going to be relaxing them. Take a deep breath in and then let it out. Three times. Inhale through your nose and blow out through your mouth."

I breathed as deeply as I could. Three times.

"Now I want you to focus on your toes. In your mind, think about your toes. Feel your toes. Wiggle them around a little. Curl them up in your shoes and then relax them. Take a deep breath in, and then out again slowly. Just keep breathing and focusing on your toes. Feel them getting heavier and heavier."

I took some more deep breaths in and out and tried to focus on my toes. You would think that would be easy, but it wasn't. I wiggled them around a little in my shoes, but then I started wondering if I would get new shoes before school started, and I started to think about not having money and I forgot all about my toes.

It seemed like Ruth knew every time I started thinking about something other than my toes because each time my mind wandered off to anything other than my toes she would interrupt at that exact moment and tell me to take a deep breath again. I can't tell you how long I had to breathe and think about my toes, but it seemed like forever.

"Now I want you to take a deep breath and focus on your feet."

I was getting hungry. I was getting bored. What did my feet have to do with learning magic? It was probably getting close to lunch. Maybe she was going to starve me to death. She must have been reading my mind because I swear she knew exactly when to interrupt me.

"Bring your mind back to your feet."

I rotated my ankles and thought about my big, stupid, hungry feet.

"Now think about your ankles. Your knees. Relax your thighs. Feel your legs getting heavy and dropping into the chair."

I imagined I was the fattest man in the world and that the chair would get so heavy it would drop through the shag carpet and end up in China.

"Now relax the muscles in your stomach. Tighten them and then relax them." I did as she said only to have my stomach growl so loud I was sure she could hear it.

"Now your chest, Jim. Take a deep breath in and out and relax your chest. Feel your heart beating and relax the muscles around your heart. Your heart is a muscle, pumping blood and oxygen through the body. You can relax it like any other muscle." I wondered if I relaxed my heart whether my body would just stop working. What would Ruth do then?

"Focus on the center of your chest. Feel the muscles of your chest relaxing. Take a deep breath in and feel your heart beating as you relax further. Now breathe out and again focus on relaxing the muscles of your chest." I noticed as I did the exercise my heart was no longer racing.

In medical school I would study the heart. I would learn that there are nerves that connect the heart to that part of the brainstem called the medulla oblongata via the vagus nerve, how the vagus nerve had two components, and how if you increased the output of the nerve by relaxing and slowing the breath, it would stimulate the parasympathetic nervous system, slowing your heart and decreasing your blood pressure. I also learned how decreasing the tone of the vagus nerve actually stimulates the sympathetic nervous system, which is what happens if you are scared or frightened—your heartbeat increases. But in the magic shop that day all I knew was that when Ruth was teaching me how to relax and breathe I felt a little better, a little calmer. I didn't know about the nervous system and the myriad ways the brain and heart communicate. Neither my brain nor my heart needed to study anything in order for it to work. I was sending signals from my brain to my heart, and my heart was responding.

"Now I want you to relax your shoulders. Your neck. Your jaw. Let your tongue drop to the bottom of your mouth. Feel your eyes and your forehead tighten and relax. Let everything, every muscle in your body . . . just . . . relax."

Ruth didn't say anything else for what seemed like forever. I sat there trying to relax, trying to breathe slowly in and out. Trying not to fidget. I could hear her taking in deep breaths and blowing them out, and I took this as a signal I should do the same. It's hard to breathe when you're thinking about how you should breathe. Once or twice I tried to peek out at Ruth through squinted eyes, and I could see she had her own eyes closed and was mirroring my position in the chair. Finally she spoke.

"OK. Time is up. Open your eyes."

I opened my eyes and sat up in the chair. My body did feel different and a little strange.

"That's it, Jim. I bet you could use a snack." She pulled open a drawer in the desk and pulled out a bag of Chips Ahoy! chocolate chip cookies and said, "Take as many as you want." I took a handful. They were my favorite. Then she looked at me over the rim of her glasses that she had put on and said, "You're on your way."

I didn't really know what I was on my way to. I wasn't sure what was really magic about just sitting in a chair for an hour.

"Jim, I want you to practice relaxing your body. Especially in situations with your family, like you told me about. You can stay relaxed even when you're feeling angry or sad. I know it seems like a lot of work, but eventually you'll be able to get into a state of total relaxation almost instantaneously. It's a great trick to learn. Trust me on this."

"OK. But can I ask why?"

"There are a lot of things in life we can't control. It's hard, especially when you're a child, to feel like you have control over anything. Like you can change anything. But you can control your body and you can control your mind. That might not sound like a lot, but it's very powerful. It can change everything."

"I don't know."

"You will. Keep coming back. Keep practicing everything you learn this summer, and someday you will."

I nodded yes, but I didn't know if I would come back or not. This wasn't like the magic tricks I wanted to learn.

"Do you know who Isaac Newton is?" she asked.

"Some kind of scientist?"

"Yes, very good. He was a physicist and a mathematician. Maybe one of the greatest scientists of all time. There's a story about him you might like. He didn't have a great life. His father died three months before he was born. He was premature and without a father, so you could say he didn't really get a fair start in life. His mother remarried when he was three years old, and he didn't care much for his stepfather. At one point he threatened to burn down the house with both of them inside. Isaac was a pretty angry young man when he was about your age. Anyway, his mom took him out of school because she wanted him to be a farmer. That's what his father had been, and that's what everyone expected him to be as well. But Isaac hated farming. He hated everything about it. A teacher convinced his mother to let him go back to school. He became the top student, but only because he was horribly picked on and bullied by other students, and getting the very best grades was his form of revenge. Later he went to college, but in order to pay for it he had to be the valet at the school in exchange for his tuition and food. He may not have had the same advantages as other kids, or the same luck, or the same money. But he changed the world."

I never knew famous scientists hated their parents or fought with their classmates.

I said good-bye to Ruth and to Neil and was just about out the door of the magic shop when I heard Ruth say, "Don't forget, Jim, practice what we talked about." She looked me directly in the eyes and smiled. I pedaled out to Avenue I with a feeling of warmth throughout my body. I had no idea why she was teaching me to relax my body, but I would go home and practice and see if it really was magic.

Today I know that a large part of what Ruth began to teach me that first day had to do with the brain and the body's acute response to stress, or what most people call the fight-or-flight response. If the brain perceives a threat, or is in fear for its survival, that part of the autonomic nervous system called the sympathetic nervous system kicks in and releases epinephrine. The adrenal gland also gets triggered by hormones released by the hypothalamus, and cortisol is produced. I'm sure even at the age of twelve I had elevated cortisol levels. Basically everything in the body not necessary for fighting for your life shuts down. Digestion slows, blood vessels constrict (except for those in your large muscles, which dilate), your hearing lessens, your vision narrows, your heart rate goes up, and your mouth gets dry because the lacrimal gland that regulates salivation immediately gets inhibited.

All of this is important if you are in fact fighting for your life, but this acute stress response is meant to be temporary. Living in a state of prolonged stress has all sorts of psychological and physiological repercussions—anger, depression, anxiety, chest pain, headaches, insomnia, and a suppressed immune system.

Long before people were talking about stress hormones, Ruth was teaching me to regulate my physiological response to chronic stress and threat. Today when I go into an operating room, I can slow down my breathing, regulate my blood pressure, and keep my heart rate low. When I am looking through a microscope and operating within the most delicate parts of the brain, my hands are steady and my body is relaxed because of what Ruth taught me in the magic shop. In fact, if it weren't for Ruth, I may not have become a neurosurgeon. Learning to relax the body is and was powerful, but it was only the start. It took ten days for Ruth to get me to a place where I could relax my entire body. On the eleventh day I rode my bike to the shop, sat in the chair, closed my eyes, and waited for Ruth to talk me through the relaxation process. But Ruth had other plans.

"Open your eyes, Jim. It's time to do something about all those voices in your head."

Ruth's Trick #1

Relaxing the Body

1. Find a time and a place to do this exercise so that you will not be interrupted.

2. Do not start if you are already stressed, have other matters distracting you, have drunk alcohol, used recreational drugs, or are tired.

3. Before beginning sit for a few minutes and just relax. Think of what you wish to accomplish with this exercise. Define your intention.

4. Now close your eyes.

5. Begin by taking three deep breaths in through your nose and slowly out through your mouth. Repeat until you get used to this type of breathing so that the breathing itself is not distracting you.

6. Once you feel comfortable breathing in this manner, specifically think about how you are sitting and imagine that you are looking at yourself.

7. Now begin focusing on your toes and relax them. Now focus on your feet, relaxing your muscles. Imagine them almost melting away as you continue to breathe in and out. Only focus on your toes and feet. When you

begin, it will be easy to be distracted or to have your thoughts wander. When this happens, simply begin again, relaxing the muscles of your toes and feet.

8. Once you have been able to relax your toes and feet, extend the exercise upward, relaxing your calves and thighs.

9. Then relax the muscles of your abdomen and chest.

10. Next think of your spine and relax the muscles all along your spine and up to your shoulders and your neck.

11. Finally relax the muscles of your face and your scalp.

12. As you are able to extend the relaxation of the muscles of your body, notice that there is a calmness overcoming you. That you feel good. At this point, it is not unusual to feel sleepy or even fall asleep. That's OK. It may take several attempts to get to this point and be able to hold this feeling of being relaxed without falling asleep. Be patient. Be kind to yourself.

13. Now focus on your heart and think of relaxing your heart muscle as you slowly breathe in and out. You will find that your heartbeat will slow as your body relaxes and your breath slows.

14. Imagine your body, now completely relaxed, and experience the sense of simply being as you slowly breathe in and out. Feel the sense of warmth. Many will feel that

they are floating and will be overcome with a sense of calmness. Continue to slowly breathe in and slowly exhale out.

15. With intention remember this sense of relaxation, calmness, and warmth.

16. Now slowly open your eyes. Sit for a few minutes with your eyes open and just be with no other intention or thought.

Breath and relaxation are the first steps toward taming the mind.

*You can visit intothemagicshop.com to listen to an audio version of this exercise.

THREE

Thinking About Thinking

A good magician signals to the audience that he's about to do his next trick. A great magician already has the audience under his spell before they even realize he's moved on to the next trick.

Ruth was a great magician.

I never knew there were voices in my head until Ruth pointed them out. I never knew how loud they were until Ruth asked me to try to keep them silent. It was hard to train my body to relax—especially at home in a small apartment where the television always seemed to be blaring and every deep breath was infused with the stale cigarette smoke that hung heavy in the air. But if relaxing my body was difficult, silencing my thoughts seemed impossible.

I had been coming to the magic shop for ten days and in many ways it was more comfortable than my own house. I loved the quiet and the calm. After the first few days of lessons, Ruth started to bring lunch every day. We would finish our magic practice and go to the front of the store and out would come a big green Tupperware container with a white plastic lid inside of which was usually sliced pieces of fruit, cheese and crackers, or nuts. The only nuts I usually liked to eat were Corn Nuts, but I tried Ruth's varieties even though some of them were weird. This was always followed by my favorite, Chips Ahoy! cookies. If Neil wasn't busy, he would join us and tell stories or show me a new magic trick or the latest card deck he was making. Neil liked to talk with his mouth full. Even though we were an odd and temporary trio, I quickly felt close to them. Sort of like they were family. I didn't have to be the caretaker in my magic shop family, and for two hours a day I had their undivided attention. We talked and joked and there was an ease about it, unlike at home where certain topics were avoided and underlying anger or resentment could surface at any time. Neil began every story by putting his reading glasses on and then looking over them and smiling at you as he began.

Neil told a story about being stationed in the Korean Demilitarized Zone. He said he and his buddies were performing a magic act inside their canteen when their commanding officer came in and demanded they immediately report to the

38th parallel—the dividing line between North and South Korea. He and his two army buddies arrived at the checkpoint, but the military police wouldn't clear them to enter because, while they had their weapons, they were still wearing the top hats and long tails from their magic show performance. I don't know if this story or any of the stories Neil told me were true or exaggerated, but they made us laugh. The kind of laughing where once you start you just can't stop. In those moments I could completely relax and let go of the voice in my head that Ruth was telling me about. Ruth told me tales of her living in a small town in Ohio where everyone cared for each other and where long summer days were spent with family and friends. Sometimes I imagined Neil taking me on as an apprentice and teaching me all his most top secret magic tricks. I could even imagine the marquee advertising the two of us in big lights. It's funny how when you've been starved of such experiences you want to hold them and not let them go. The connection I had with Ruth and Neil was special and real. I've felt that connection with others throughout my life—sometimes it's a random person in an elevator, where you look into each other's eyes, and for reasons you can't explain, there is a connection, not just simply eyes meeting, but some deeper knowing, an acknowledgment of each other's humanity and the reality of being on the same path. And when that happens, it's pretty magical if you really think about it. Other times I've looked into the eyes of someone who is homeless or just down-

and-out, and when our eyes connect it is as if I could see my very own face staring back at me and for that brief moment, and often even longer, I experience the pain of my own journey and feel deep empathy followed by gratitude that my journey has taken me to where I am today. Everyone has a story, and I have learned that, at the core of it, most of our stories are more similar than not. Connection can be powerful. Sometimes just a brief meeting can change someone's life forever.

Clearly that was the case with Ruth. That first encounter changed everything, putting my life on a far different trajectory than it would have been on. Ruth wasn't a supernatural being, even though at twelve I liked to imagine she was. She was simply a human being who had the profound gift of empathy and intuition, of being able to care about another human being without expecting anything in return. She gave me her time. She gave me her attention. And she exposed me to a type of magic that I still use to this day. There were some hours in the magic shop where I was convinced that being there was a waste of time and that I couldn't possibly learn what she was trying to teach me. There were other times when I truly thought she was pretty close to crazy. Today I know that the techniques Ruth was teaching me were in many ways age-old and had been part of Eastern traditions dating back thousands of years. Now science acknowledges that neuroplasticity is not only a reality but an inherent part of how the brain functions. Now I know that the brain can be trained to improve one's focus and attention and also to not respond to the ongoing dialogue in our head that distracts us from making clear and useful decisions. Today this is well understood, but at that time what Ruth was teaching me was unheard of. When Ruth told me she was going to teach me to turn off the voices in my head, I had no idea what she was talking about, but I decided to go along with it anyway.

"Relax your shoulders. Relax your neck. Relax your jaw. Feel the muscles in your face relax," she said, all things that I now knew how to do.

Ruth talked me through relaxing my body yet again, her soft voice making my body feel so light I wouldn't have been surprised if I was hovering over the chair, levitating like a playing card out of one of Neil's magic rising decks.

"Now I want you to empty your mind."

That was a new one. I suddenly felt the weight of my body against the chair. What was Ruth talking about exactly? How was I supposed to empty my mind? My thoughts were off and running and I opened my eyes to see Ruth smiling at me.

"This is another trick," she said.

"OK. How do I do it?"

"Well, this gets a little complicated because your mind is going to think about thinking, and the minute it does that you're going to have to stop thinking about thinking without thinking about it."

Huh?

"Do you know what a narrator is?"

"Sure," I said. "It's like you guiding me through the relaxation trick."

Ruth clapped her hands together twice and laughed a little. "When you do the relaxation trick at home, how do you do it?"

I thought about this for a second. "I do it the same way I do it here."

"Well, I'm not there narrating, so who narrates it?"

"You do, but in my head."

"But it's not really me in your head, so who is narrating it?"

As far as I was concerned, it was her voice in my head telling me to focus on and relax every muscle in my body. "It's your voice."

"But it's not actually me, so who is it?"

I guessed at what she wanted me to say. "It's me?"

"Yes, it's you, talking to yourself in your head, and it sounds like me because that's what you want it to sound like. This narrator is very good at doing impersonations. It can sound like anyone."

"OK."

"We all have this voice that talks to us nonstop in our heads. From the minute we wake up to the minute we go to bed at night. It's always there. Think about it. It's like one of those radio deejays telling you what's coming next. Giving you the playlist every second of the day."

I thought about this. I listened to Boss radio, Top Forty hits, 930 on the KHJ-AM dial in Los Angeles. I imagined "the Real Don Steele" narrating my life.

"Imagine this deejay in your head telling you everything about everything all day long. You're so used to it you probably don't even notice that the radio in your mind is playing at full volume, and it never gets turned off."

Was this true? I wasn't sure. I hadn't noticed it before. I was always thinking about stuff, but I had never really thought about thinking before.

"This voice in your head is judging every second of your life as either good or bad. And your mind responds to what the voice is telling you. As if it *actually* knew you." Ruth said this with emphasis, as if I should be shocked or affronted by me thinking about me. I was totally confused. "The problem is that often your response isn't one that is necessarily good for you."

"Well, it's me in my head, so don't I know me?"

"No. You are not the voice in your head. You, the real you, is the person who is listening to the deejay."

I wondered just how many people Ruth thought lived inside me. Maybe she heard voices in her head, but I was pretty sure it was just me in my head, not some deejay telling me the weather and cuing up the next song.

"Here's what I want you to understand. You can't trust the voice in your head—the one that's talking to you all the time. It's more often wrong than right. You can think of this trick as learning to turn the volume way down and eventually turning it off altogether. Then you'll understand what I'm talking about."

"I guess I could give it a try," I said to Ruth.

"What's the deejay saying right now? Right at this very second, in your head?"

I thought about what I was just thinking. "He's saying that I have no idea what you're talking about, and this isn't going to work." The deejay was also saying that this whole thing sounded really stupid, but I wasn't going to tell Ruth that.

She smiled at me. "That's good. You see, you just thought about what you were thinking about. That's the first part of this trick."

I nodded as if I understood.

"We're going to practice thinking about thinking. Now close your eyes and take a few minutes to relax your body again."

I closed my eyes and went through the relaxation sequence I had practiced a hundred times by now. I started with my toes and worked my way up to the top of my head . . . every muscle relaxing as I thought of it in my mind. By now it felt good, like being in a tub that was slowly filling up with warm water.

"Just focus on your breath," Ruth said. "In and out. Just think about your breath. Nothing else but your breath."

I took a breath in through my nose and slowly exhaled. And then another. After a few more breaths I felt an itch on my face and moved my hand up to scratch it, and as I did so I felt a bump. I hoped it wasn't a pimple coming up. There was a girl I liked who had just moved in above us in the apartment complex. Her name was Chris. Her hair was long and dark, almost down to her waist. I had spoken to her that first day I saw her and afterward wondered if she thought I was a dork. She was nice enough and smiled as we talked. Would she consider hanging out with me? I suddenly also remembered my crooked tooth and ran my upper lip over it. No, she wouldn't. What was I thinking? Pimples and a crooked tooth, geez. I remember her looking at me and then turning and walking away. I wasn't good enough for her.

"Keep focusing on your breath. If the deejay starts talking, just stop listening and go back to focusing on your breath."

My mind had gone off, and I hadn't even noticed it. I went back to thinking about my breath but then started thinking about hanging out with a guy from my class. He lived in the "nice" part of town. His father owned a construction company, and they lived in a huge house, and his parents drove matching Cadillacs. He had invited me for dinner once last year, and during dinner his mother asked where I lived, and then what type of work my father did. I wanted to crawl under the table and disappear. My father didn't have a job and had been arrested on more than one occasion for being drunk and disorderly. It wasn't something I could tell her and probably wasn't something she wanted to hear.

I had done it again. I was thinking about something other than my breathing. This was hard. I couldn't do it. It seemed like I could only take about five breaths before I started thinking about something else. I decided to count how many breaths but then realized that if I was counting breaths I was still thinking. This was actually impossible. Are people really able to do this? Could Ruth do this trick? How many breaths could she take without thinking about something? Should I ask her? Did it take Ruth a long time to learn or was I just really bad at it? What's the point anyway? And on and on I went.

I tried my best to slow down my thoughts, but apparently my mind was unable to sit still like the rest of me. *Would Ruth know if I just faked it?*

"Open your eyes."

I looked at Ruth. I had totally failed this one. "It's too hard," I said. "I can't do it."

"You can do anything, Jim."

"Not this."

"It just takes practice. Just try to stop your thoughts for a second. Then a few more seconds. Then a little longer."

"I'm really not good at this."

Ruth just looked at me and said nothing for a few seconds.

"Everyone who tries this says the same thing at first. You can be good at anything you want. Even this. You just don't know it yet."

I suddenly felt the pain of all the times that I felt that I wasn't good enough or didn't belong or couldn't afford something. And just like that I felt my eyes start to sting. Every once in a while, during that time with Ruth, those feelings would well up and I would want to lay my head down and cry.

"When your mind wanders away from your breath it's not good or bad. It's just doing what it does. Just notice it. Then guide it back to your breath. Help it focus again. That's all. You just have to show it who is in control. All I want you to do is notice when you are thinking. Then you'll begin to notice when your mind isn't running all over the place."

"I'll practice."

"Excellent. That's all you can do. Practice, practice, and more practice."

"Is that the way it was with you?" I asked.

"Exactly the same," she said. I already felt better.

"Do I relax my body first?"

"Relax first, then calm your mind by taming your thoughts. Eventually, all the tricks I teach will just flow together and you'll relax and quiet your mind at the same time, but for now do it step-by-step."

. . .

I WENT HOME that day determined to master the art of silencing the obnoxious deejay in my head. My dad was still gone when I got home, and my mom was in her room in bed. I sat in silence in my room concentrating on turning off the deejay, slowly breathing in and out, but the silence only seemed to make the voice in my head grow louder. I knew my dad was on a drinking binge and at any moment he could burst through the door either really drunk or really hungover. It was like this scene in my life was on repeat—playing out over and over again, always the same. He would walk in the door, my parents would have a loud fight, he would blame her for all his problems in the past, and then make promises for the future that he could never keep. Over and over again.

If anyone in my family noticed me sitting in a chair with my eyes closed, more often than not they never said anything about it. No one asked me what I was doing. No one asked me what I was thinking. And they certainly never asked what I was feeling. I tried my best to practice Ruth's magic, but with every day that my dad stayed away, I could only wonder and worry about what was going to happen when he finally showed up. *How would the argument begin? What if my mom overdosed on pills again?* I tried to stop thinking, but it was impossible. *Would I call the police or an ambulance? Who would I have to talk to? How would I explain my brother hiding under the covers in our*

room when they came for my mom? Would they take my dad away?
I tried to focus my mind on my breathing, but my mind could
only conjure up disaster scenario after disaster scenario—each
one beginning with my father walking through the front door.
It was like knowing there was a tornado about to touch down
but being so frozen in fear you couldn't run and take cover.
Sometimes I had dreams like that. Nightmares really. Where
I opened my mouth to scream out a warning to someone but
no sound would come out.

Ruth seemed to know I was struggling, because she switched
things up on me a few days later.

"Let's try a different way to stop all those thoughts in your
head."

Ruth had brought a candle and she lit it with a little card-
board match. She put it on the office desk. She had me move
my chair so that it was facing the candle.

"I want you to focus on the candle. The light of the candle."

She had me take deep breaths in and out and just stare at
the lit candle.

"Just think about the light. Every time your mind wanders,
focus it back on the light."

In a way, it was easier for me to quiet my mind with my eyes
open. It was when I closed my eyes and everything went dark
that most of my worries came rushing out. In the dark there
was no distraction, and every fear seemed to want to come out
and play. *When were we going to get evicted again? Why did my*

dad have to drink? Was my mom ever going to get better? When would we have money? Why couldn't I fix my family? What was wrong with me? When I stared at the candle flame it was like I could get lost in it. I could focus on the blue at the bottom of the fire, then on the orange in the middle, which looked like a Halloween candy corn. Sometimes I would focus on the white tip of the flame. It almost felt like I could go inside it. It was so much easier to quiet the deejay simply by staring at the single flame that would flicker ever so slightly with each breath I took. It also reminded me of the time when friends of my family invited us several years before to their cabin in the mountains. There was a fireplace and I remember sitting in front of it. During that brief period of time my father had a job. He hadn't gotten drunk in some time. My parents were civil, and my mother's health seemed better. I sat in front of the fire and looked at the flames and for a while I was lost in them. I was feeling warm. Feeling good. Feeling happy.

I spent so many hours over those weeks with Ruth watching that candle. To this day the sight of a lit candle brings me to a place of calm. I didn't have a candle at home that first day. I remember going with a friend to the Catholic church several weeks before because his grandmother was ill, and he put a dime in a box on the inside of the church and lit a candle and said a prayer. It seemed very foreign to me. On the way home, I made a detour to the church and took two candles and some matches, leaving the fifteen cents that I had in my pocket.

And every night, I struggled and stared at the candle flame trying to stretch the gap between my thoughts.

As a surgeon I have often heard my patients describe how they experience pain more acutely at night—it's not that their pain is worse at night, it's just that there's no distraction. The mind gets quiet and the pain that was there all day seems louder. It's the same reason why our eyes can fly open at 2 A.M., and every anxiety about the future or regret about the past will play itself out in the dark of night. Ruth taught me how to control my mind, and in doing so she helped me stop reliving the guilt and shame of past events and the anxiety and fear of imagining possible future events playing on the radio station of my mind. Or perhaps more important, she taught me not to respond emotionally to these thoughts the same way I had previously. She taught me the pointlessness of wishing for a different past and the futility of worrying about all of the frightening futures over which I had no control.

In all we spent almost three weeks practicing three different ways to make me aware of my thoughts and bring quiet to my mind. Focusing on my breath, staring at a candle flame, and the final method—chanting.

"DO YOU KNOW what a mantra is, Jim?"

I shook my head. I didn't have a clue.

"It's kind of like a song or a sound you make that helps you focus your mind. Just like you've been focusing your mind on your breathing or the candle, this is another way to trick your mind."

I looked at her again and noticed she was wearing a necklace with a whistle and a bell. Is that what she was talking about? At that moment she leaned forward toward me and the bell made a little tinkle. I almost started laughing. She looked down at it and laughed. "No, that's not what I'm talking about."

"What kind of sound?" I had a feeling this was going to be weird.

"Well, it depends. People sometimes say a word that is important to them or a phrase that has some magical meaning. But it can be anything. The words don't really matter; it's the sound that matters."

"So what do I say?" I asked.

"That's up to you. Whatever it is, you are going to chant it over and over again."

"Out loud?"

"No, to yourself."

This was definitely going to be weird. I had no idea what important words I was supposed to come up with. The only words I had ever said over and over again in my head were curse words, and I was pretty sure that wasn't what Ruth had in mind.

"So what's it going to be?" Ruth was waiting patiently for

me to come up with some magical word, and I had absolutely nothing.

"I don't know." I knew that, in magic, words were important. Abracadabra. Open sesame. These words had to be just right to work.

"What is the first word or words that come to mind? Anything at all."

"Chris," I said to myself. It was the girl from the upstairs apartment. I was searching in my head for what I thought would be an appropriate word. I couldn't think of anything else. Suddenly the image of a doorknob popped into my head. A knob. Chris knob. To this day, I don't know how I arrived at that combination of words or what meaning they had to me at that moment.

Ruth looked at me. "Well, do you have it?"

"Yes," I said, but I suddenly felt shy. I had chosen the wrong words. They were going to sound stupid and probably wouldn't work.

"Now say it to yourself, but slowly, and stretch out each word as you say it."

"Chriisss . . . Knobbb . . ." I said it to myself.

I did it again a few times in a row.

"Now I want you to chant it to yourself. Over and over for the next fifteen minutes."

Ruth looked at me and I'm sure I looked back at her like she was out of her mind.

"Just focus your mind on the sound of each word. Don't think about anything else."

Ruth was right. It was hard to think about anything else while I was chanting my made-up mantra. And even though I was saying the word *Chris* combined with the word *knob* over and over again, I couldn't even focus on her or the doorknob. It didn't matter if she knew I existed or what she thought of my tooth or if she noticed I had a pimple. That wasn't the point. The point was, I didn't hear the deejay. He had stopped playing.

I PRACTICED MY MANTRA at home. Sometimes for hours at a time. For reasons that I understand now, it was amazingly calming. Repetition. Intention. The surest way to change your brain. By combining the breathing technique that Ruth had taught me with either looking at the flame of a candle or slowly repeating my mantra, things began to change.

Eventually, my father did come home. This time he was hungover and repentant. My mother had come out of her room, and it began. The usual arguments, but this time it included the fact that we had been given an eviction notice. I had been in my room for the last few hours practicing my breathing and chanting to myself. For reasons that I can't explain, I walked into the room and told them I loved them. I realized I saw them in a different way. I went back to my room.

I didn't feel angry or upset. I accepted the situation. I realized after a few minutes that I didn't hear anything either in my head or outside of it. The house had gone silent. I walked back out to the living room and saw that my parents were just sitting there quietly.

"It's going to be OK," my dad said.

"We love you too," added my mom.

At that moment, I didn't really know if things were going to be OK or not. I knew they loved me as best they could. And that was far different from how I had hoped for so long that they would love me. Yet at that moment, it felt like enough.

THE FIRST BRAIN I ever saw was suspended in a glass jar full of formaldehyde. It was gray and furrowed—more like a giant walnut or a three-pound lump of old hamburger than a supercomputer responsible for all human functioning. I stared at the wrinkled mass, and my mind wondered how such a gelatinous blob of gray and white matter could be the source of thought, language, and memory. I would learn the places in the brain responsible for speech and taste and all motor functions, but there was no instructor who could ever show me—not in a textbook or during surgery—what part of the brain I could slice into and watch love spill out. There was no cross section that would show a mother's drive to nurture and protect her child. There was no small sliver I could biopsy that held the mysterious force that could make a father work two jobs just so his children had more than he had growing up. There was no tangible center in the brain that I could pinpoint as the place that caused one person to rush to the aid of another person—or strangers to come together in times of crisis.

What part of the brain was it exactly that had made Ruth want to give me her time and attention and love?

I couldn't see any of these things in a brain floating in formaldehyde, and I couldn't see them through a microscope while performing brain surgery. I spent many late nights during medical school using my brain to think about the brain and then using my mind to ponder the irony of it. How exactly do we separate and distinguish the mind from the brain? I can operate on the brain but not the mind, but operating on the brain can forever alter the mind. It's a dilemma of causality—a circular reference problem like the perennial question of what came first, the chicken or the egg. One day I asked Ruth this very question.

"Jim," she said, "if you're hungry, it really doesn't matter whether the chicken or the egg comes first, does it?" I had at times been very hungry, and I would have happily eaten a chicken or an egg.

She always had a way of breaking things down and putting them in perspective. And day after day, she was teaching me how to get a new perspective on my own feelings and thoughts. And this thinking about thinking—this ability of the brain to observe itself—is one of its great mysteries.

With only two weeks left in our summer together, and just as I was wrapping my mind around the idea that I could observe my thoughts and therefore I was separate from my thoughts, Ruth pulled a whole new trick out of her bag.

"Jim," she said, "have you seen the trick where the magician saws a woman right in half?"

I nodded. "Of course."

"Well, we're going to do a trick kind of like that, but with your heart. We're going to cut it open. Split it right down the middle."

I had no idea what she was talking about, but by this time I was used to Ruth springing things on me, and I knew all I could do was settle in, buckle up, and enjoy the ride.

Ruth's Trick #2

Taming the Mind

1. Once your body is relaxed (Ruth's Trick #1), it is time to tame the mind.

2. Begin again by focusing on your breath. It is common for thoughts to arise and for you to want to attend to them. Each time this occurs, return your focus to your breathing. Some find that actually thinking of their nostrils and the air entering and exiting helps bring their focus back.

3. Other techniques that assist in decreasing mind wandering are the use of a mantra, a word or phrase that is repeated over and over, and focusing on the flame of a candle or on another object. This helps avoid giving those wandering thoughts attention. In some practices, the teacher gives the mantra to the student who tells no one else the mantra, but you can pick whatever word you like as your mantra. Or you can focus on a flame or on another object. Find what works best for you. Everyone is different.

4. It will take time and effort. Don't be discouraged. It may take a few weeks or even longer before you start

seeing the profound effects of a quiet mind. You won't have the same desire to engage emotionally in thoughts that often are negative or distracting. The calmness you felt from simply relaxing will increase because when you are not distracted by internal dialogue the associated emotional response does not occur. It is this response that has an effect on the rest of your body.

5. Practice this exercise for twenty to thirty minutes per day.

*The reward for taming the mind
is clarity of thought.*

*You can visit intothemagicshop.com to listen to an audio version of this exercise.

FOUR

Growing Pains

I left earlier than usual for the magic shop because it was expected to be one of the hottest August days on record in Lancaster—triple digits. The sky was full of wispy clouds that looked more sooty than white. It wasn't sunny and it wasn't cloudy, and everywhere you looked was either brown or gray. I could feel the heat coming up from the ground through the pedals on my bike, so hot I thought it would singe the hair on my legs. I had to alternate one hand at a time on the handlebars so both hands didn't feel like they were burning. I tried riding no-handed for a while down Avenue K and was just getting up a good rhythm when I heard yelling from the field next to the Episcopal church.

I recognized the bigger kid, the one who was throwing the punches. He was two grades above me, and both my brother and I had been pushed around, hit a few times, and even spat on by him and his trusty sidekick. They were a gang of two and pretty much ruled Lancaster in the afternoon between the hours of three and five during the school year. Obviously they were operating on extended summer hours because here it was not even 10 A.M. and I could see one of them punching and kicking a kid while the other yelled and laughed. I couldn't see who it was because the kid on the ground was curled up and had his head down. His arms were wrapped around the top of his head trying to protect it. For a second I thought it might be my brother but then remembered he had actually been home when I left.

I'm not sure what it was that made me get off my bike and start yelling at the boys. I was used to defending my brother, a habit that I would carry with me into adulthood, but I didn't go looking for fights, and certainly not with these guys. They didn't hear me at first, and as I walked toward them, it was like I could feel every punch and kick they delivered to the boy on the ground and my heart started to hammer in my chest. I took a deep breath and yelled again for them to cut it out.

"Stop it!"

The big guy was hunched over the kid, and when he heard me, he stood up tall. He gave me a snarly grin and then kicked the kid on the ground one more time in the stomach. It made me flinch and feel like I had just been kicked in the stomach myself.

"Who's going to make me?"

Their attention diverted to me and I saw the kid on the ground roll onto his back and start to get up. It was a kid I kind of knew from school. I couldn't remember his name, but I knew his family had transferred here last year. His dad was out at the air base. The kid's face was bloody, and his glasses were in the dirt next to him. He had to be half the size of all three of us. I was as tall as these older kids, but they outweighed me by at least thirty pounds. I watched as he got to his feet and started staggering toward the church. I couldn't blame him for getting the hell out of there.

"You going to take his place?"

The two boys took a few steps toward me, and I felt my mouth go dry and my ears start to buzz. I tried taking some deep breaths the way Ruth had taught me, but I couldn't seem to get the air to fill up my lungs.

This was not going to be good.

"So, you think you're a hero? Some kind of freaking hero?"

I didn't say anything. I tried relaxing my legs and my hands like I had learned in the magic shop. I bounced up and down on the balls of my feet and cleared out my thoughts. If I had to fight, I would. I wasn't going to run.

"I'm going to kick your ass and then we're going to take your bike."

I still didn't say anything. I sensed the sidekick moving behind me a little, but I just stared straight at the guy who liked to punch and kick. He was the one who called the shots for the pair. He moved his face so close to my face that I could see some sort of white gunk in the corner of his mouth. It was getting hotter out by the second and his face was sweaty and dirt-stained.

"Unless you want to kiss my feet."

I thought of Ruth and Neil in the magic shop. They would be waiting for me to ride up right about now. Would Ruth think I had skipped a day with her when I didn't show up? Would anyone find me out here bleeding? Did the other kid go to get help? Did this guy wake up, have his cereal and milk, and just run out of the house ready to beat people up all without even wiping his mouth? All these thoughts started racing through my mind, but I just stared at the dried white gunk and pretended it was the light on a candle.

"Kiss my feet."

I looked up and into his eyes and spoke for the first time since I had told him to stop beating the other kid.

"No."

He reached out and grabbed the front of my T-shirt.

"Kiss my feet," he threatened. His mouth began to make a smile like someone who knows he has power over another. His face got right up to mine and I could smell and feel his breath. I closed my eyes for just a second and in that second something was different.

I opened my eyes and looked directly into his. I stared deep into his eyes, the way we do when we're trying to really understand something or someone. "You can do anything you want to me, but I'm not kissing your feet."

He laughed and looked to the side at his friend. I saw him raise his eyebrows and then he looked back at me. I stared at him, without blinking. He lifted his fist and cocked it behind his ear. I didn't flinch. I just kept my eyes locked on his and in that moment I didn't care that he was bigger than me or that there was some other kid's blood on his fist. I wasn't going to back down. I wasn't going to give him the power to make me afraid. And I wasn't going to kiss his feet or anyone's feet. Ever.

And for a second our eyes locked together and I saw him, and he knew I saw him. I saw his own pain and fear. A pain and fear that he tried to hide with his bullying.

His gaze broke from mine and he looked at his sidekick and back at me. "What a waste."

He let go of my shirt and pushed me a little so that I stumbled back a step but I didn't fall down.

He didn't look at me again for the briefest of moments and turned away. "It's too hot. Let's get out of here."

I felt the other kid give a little push against my back, but it was more for show than anything else. I could tell he wasn't sure what had just happened. They both started walking away, and I could see the other boy talking to the bully. I knew he was asking why he didn't beat me up. The bully pushed him and said, "Shut up." Neither of them looked back.

I took a few more deep breaths and watched them as they walked away before I turned toward my bike. I wasn't exactly sure what happened or even why I did what I did but I felt good. Suddenly I realized I was late and Ruth was waiting for me. I hoped she didn't think I had just blown her off. I got on my bike and raced as fast as I could to the magic shop.

I WENT bursting through the door, out of breath but ready to tell Ruth and Neil the whole story of what had happened on my way to the shop. I had stood up for myself and had stood up for a little kid who couldn't defend himself. For probably the first time, I felt like a hero. Ruth *had* to forgive me for being late once she realized what I'd done.

"Ruth," I called out. It was strange; neither she nor Neil were at the counter. "Ruth! Neil! I'm here."

Nothing.

I headed back toward the office and that's when I heard their voices. Ruth and Neil were arguing. I had never heard them argue.

"He's just a boy."

"He's going to remember this for the rest of his life. You have to make it right."

"It's too late. The damage has been done. I'll explain it all to him when he's older."

"Damage can and should be undone." Ruth sounded angry.

I had never heard her sound like that and it worried me. Had I done something wrong? Were they that mad about me being late? None of it was making sense. What damage had Neil done to me? What was he going to explain to me when I was older?

"Neil, everybody makes mistakes. I certainly made my share with you. But I'm telling you it's not too late to fix this. You'll regret it if you don't. Trust me."

Everything got quiet. I didn't want them to walk out and see me eavesdropping. I walked back to the front of the shop and opened the door again and called out their names. Maybe they wouldn't know I had overheard them.

"Hello," I called out. "Ruth, I'm here."

Ruth walked through the office door. Her eyes were red like my mom's, so I knew she had been crying.

"Jim," she said, "you're late."

"I'm sorry. I had a little problem on my way here."

Ruth looked me up and down. "Is that blood on your shirt?"

"Yes," I answered, "but it's not mine. Don't worry."

Ruth laughed. "That worries me even more. Come on back."

I walked past Neil, and he mumbled hello but didn't look at me. I wasn't sure what I had done or what he had done, but it must have been bad. It seemed like he hated me now.

Ruth had me sit down in the chair and walked me through the relaxation exercise and then asked me to chant my mantra in my head. I started, but I couldn't stop replaying the conversation I had overheard. What mistake had Neil made with me? What was so bad that Ruth would be crying? I couldn't take it anymore, and I certainly couldn't tame my thoughts right now.

"What happened? What did I do? Why is Neil mad at me?" I blurted out all three questions with my eyes still closed and then I opened them to see Ruth looking at me with a puzzled look.

"Why would you think you did something?" she asked.

"I heard you and Neil arguing about me. I heard you through the door. He hates me."

Ruth continued to stare at me and then she just nodded her head.

"You heard all that?"

"Yes," I said, miserably. I knew Ruth and Neil had been too good to be true, and I was pretty sure this was my last day at the magic shop.

"Really, now? And what did Neil say about you?"

"He said . . ."

I thought about it but couldn't remember exactly what Neil had said about me.

"Yes?" Ruth prompted.

"It was something about . . . something about the damage being done."

"And you heard your name?"

"No, not exactly," I said. I couldn't remember them saying

my name, but I knew it was about me. I felt even more miserable. Was Ruth going to lie to me and tell me they weren't fighting about me?

"Jim," Ruth said gently, "we weren't talking about you. We were talking about my grandson."

"Your grandson?"

"Yes, Neil has a son, and it's complicated and sad, and I miss him."

"How old is he?"

"He's around your age."

"Where is he?"

"He's with his mom right now. But that's not important. What's important is why you thought our argument was about you. Why you would think Neil hated you."

I didn't really know what to say to that. I had just assumed they were talking about me.

"Jim, everyone has situations in their life that cause them pain. The situation with my grandson and my son hurts my heart. It's like a wound. Now, if I cut open my knee what am I to do? I can give it some attention—clean it off, bandage it up, and make sure it heals properly—or I can ignore it and pretend it's not there, pretend it doesn't hurt or sting and just pull my pant leg down over it and hope it goes away. Is that the best way to heal it?"

"No."

Once again I wasn't sure exactly what she was talking about.

"It's the same with the wounds in our heart. We need to give them our attention so that they can heal. Otherwise the wound continues to cause us pain. Sometimes for a very long time. We're all going to get hurt. That's just the way it is. But here's the trick about the things that hurt us and cause us pain—they also serve an amazing purpose. When our hearts are wounded that's when they open. We grow through pain. We grow through difficult situations. That's why you have to embrace each and every difficult thing in your life. I feel sorry for people who have no problems. Who never have to go through anything difficult. They miss out on the gift. They miss out on the magic."

I nodded at Ruth. I had spent a lot of my life so far comparing myself to my friends who seemed like they had everything. They didn't have to stand in line at the grocery store and feel the pain as the cashier looked at you when your mom handed her food stamps. Or to wait in line at the government food bank for someone to give you a handout of powdered milk, butter, and a bland white block of cheese. They didn't have parents who argued, got drunk, or overdosed on pills. They didn't go to bed at night feeling like everything wrong was somehow their fault. They had cars and money and clothes and girlfriends and nice houses to live in. Ruth felt sorry for them?

"Jim, the next trick I'm going to teach you is to open your

heart. Some people have a lot of trouble with this. For you, it's going to be easier."

"Why?" I asked.

"Because life has already begun to open your heart. You care, Jim. You care for your family. Your brother, your mother, and even your dad. You cared when you thought Neil was mad at you. You care enough to come here every day. I have no doubt about your ability to care about others—that's part of opening your heart."

I thought about the boy who had been getting beat up that morning. I didn't really know him, but I did care. I cared enough to stop my bike. I knew I cared because I could have been (and had been) that kid. I cared because I had felt pain and humiliation a million times already and it hurt. It hurt a lot.

"The other part of opening your heart, and this is where you are going to have to really practice, is caring about yourself."

I cared about myself. That was going to be easy.

"There's a reason why you assumed our conversation was about you, Jim. You made a big leap from what you heard to Neil hating you."

"I just misunderstood," I said.

"Yes." Ruth laughed. "We all misunderstand. Each other. Ourselves. Situations. It's a good lesson to learn—that not everything is about us. I think I need to learn that same lesson when it comes to my grandson."

I nodded.

"Each of us chooses what is acceptable in our lives. As kids, we don't get a lot of choice. We are born into families and situations, and it's all really out of our control. But as we get older, we choose. Consciously or unconsciously, we decide how we are going to allow ourselves to be treated. What will you accept? What won't you accept? You're going to have to choose, and you're going to have to stand up for yourself. No one else can do it for you."

I NEVER GOT a chance to tell Ruth about the first fight I witnessed that morning, and I never heard Neil and her fight again. Every day for the next week, she taught me to open my heart. She explained to me that so often the conversation going on in all of our heads is one that is hypercritical and negative. One that frequently causes us to react in a way that isn't in our best interest. One that causes us to relive events over and over or to wish for things that might be or should be. So much so that we aren't really here a lot of the time. We started that morning with Ruth having me say nice things to myself. How strange. Over and over I said repeatedly, *I am good*, *it's not my fault*, *I'm a good person*. It was like I was another deejay at the radio station, but everything I said was nice and comforting. Every time I caught myself listening to the other deejay, I stopped and began the kindness mantra to myself.

"I am worthy. I am loved. I am cared for. I care for others. I choose only good for myself. I choose only good for others. I love myself. I love others. I open my heart. My heart is open."

Ruth asked me to make a playlist of these ten affirmations and repeat them every morning, every night, and just anytime if it popped in my head, and especially after I did my relaxation exercise and tamed my thoughts. They were all kind of hokey, but I went along with it and was grateful she hadn't asked me to say them out loud. Next, she told me she wanted me to send loving thoughts to myself, my family, my friends, and even people I didn't like or who didn't deserve it. She saw me look confused when she said to send loving thoughts to those I didn't like or those who didn't deserve it. She looked at me with a deep kindness and said, "Jim, oftentimes those who hurt people are those who hurt the most." But it was hard. It was hard to think of the bully who had beat me up and somehow think it was OK. It wasn't and I still hated him and all the other people who had been mean to me and hurt me. But I kept trying. Over and over. And after a while I found that if I thought of them being hurt or being beat up and crying in pain and then what it felt like when it happened to me, it was easier. Easier when I began to realize that when I was angry with someone, it was usually because I was hurting on the inside. I was angry at myself for something. I had never realized that before. Her words kept coming back to me: "Those who hurt people are often those who hurt the most." She was right.

And that was her point. If you can heal your own wounds, you don't hurt anymore and you don't hurt others. Wow. Was being with Ruth somehow healing me?

The week before, Ruth had told me that the last thing she was going to teach me was the power to get anything I wanted. I was ready to move on to that. I was getting a bit tired of talking about the heart. A lot of the time thinking about it made me hurt. It brought up so many painful things that I had spent a lot of time trying to bury deep down inside so they wouldn't hurt so much. But I was finding that while it really hurt when they came up, each time it was easier and not quite as painful. And finally, while I could relive the event in my mind, the emotional response wasn't quite the same. I could sit with it and not get lost in the hurt and the pain. I could sit with it and not blame myself or somehow think it was my fault. I could just be with it. I was finding that while the deejay was still there, I just wasn't paying as much attention or the sound had gone down really, really low.

Ruth was sawing my heart wide-open, and while it hurt at times, it also felt good.

ONE THING every human has in common is the first sound we hear. It's the heartbeat of our mother. That steady rhythm is the first connection each of us knows, not with our minds, but the knowing is there in our hearts. The heart is where we find our comfort and our safety in the darkest of places. It is what binds us together and what breaks when we are apart. The heart has its own kind of magic—love.

When Richard Davidson, at the University of Wisconsin, first began studying compassion, it was with Tibetan monks who were long-term meditators. The monks were told they were to wear a cap on their heads, and this cap would be embedded with innumerable electroencephalogram (EEG) electrodes to measure their compassion. When the monks heard this, they all began laughing. The researchers thought it was because the cap looked funny, with all the electrodes, each connected to a long trailing wire so that the cap resembled a wild wig. The laughter of the monks wasn't because of the cap, as the scientists thought. The researchers had it all wrong. A monk finally explained what they had found so funny. "Everyone knows," he said, "compassion doesn't arise from the brain. It comes from the heart."

Research shows the heart to be an organ of intelligence, with its own profound influence not only from our brain but on our brain, our emotions, our reasoning, and our choices. Rather than passively waiting for instructions from the brain, the heart not only thinks for itself but sends out signals to the rest of the body. The part of the vagus nerve that arises in the brainstem and that has immense innervation in the heart and other organs is part of the autonomic nervous system (ANS).

The pattern of heart rhythms known as heart rate variability (HRV) is a reflection of our inner emotional state and is influenced by the ANS. In times of stress or fear, the vagus nerve tone decreases and there is a predominance of expression of that part of the ANS called the sympathetic nervous system (SNS).

The SNS is associated with a very primitive part of our nervous system designed to respond to threat or fear by increasing blood pressure and heart rate as well as decreasing heart rate variability. Conversely, when one is calm, open, and relaxed, the tone of the vagus nerve is increased and the expression of the parasympathetic nervous system (PSNS) predominates. The PSNS stimulates our rest-and-digest response, while SNS stimulates the fight-or-flight response. By measuring HRV, researchers are able to analyze how the heart and nervous system respond to stress and emotions. Feelings of love and compassion are associated with an increase in HRV, and when we feel insecurity, anger, or frustration, our HRV decreases, becoming more smooth and regular. Many people get this confused because it would seem logical that with an increase in stress and heart rate, our HRV should become chaotic, irregular, and highly variable. And, vice versa, when the HRV is steadiest is when we should be the most calm and relaxed. HRV, however, is just the opposite of what we expect.

Interestingly, one of the greatest causes of sudden cardiac death is lack of heart rate variability—a result of chronic arousal to threat and decreased vagal nerve tone. Stress, anxiety, chronic fear, negative thinking can all cause blood to pound into the heart with extra force. It's the body's equivalent of screaming "Fire!" in a crowded theater. Over and over again. Eventually, somebody is going to get trampled.

Ruth was helping me form new neural connections in my brain. It was my first experience with neuroplasticity, well before the term was commonly used. In fact, although American psychologist William James first presented the theory over 120 years ago, it wasn't until the later part of the twentieth century that it became understood that neuroplasticity was even possible. Not only was Ruth training me to change my brain by creating new neural circuits but she was also training me to regulate the tone of my vagus nerve and, by doing so, affect both my emotional state and my heart rate and blood pressure. With only an intuitive sense of the effect of what she was teaching me and knowing none of the physiology behind the magic, she was making me more focused and attentive, calmer, boosting my immune system, lowering my stress, and even lowering my blood pressure. My mother asked me one day if I was using drugs. To that point, I had never done so. I was terrified of alcohol and drugs. By this time my mother had attempted suicide with drugs several times. She told me I seemed much calmer and happier. She told me I didn't seem as on edge. Ruth was improving my ability to regulate my emotions, increasing my empathy, my social connectedness, and making me more optimistic. She changed how I perceived myself and how I perceived the world.

And that changed absolutely everything.

THE BEST and most skilled magicians know how to control the attention of an audience, manipulate its memories, and influence its choices without the audience having a clue this is what's going on. By teaching me to relax my body and tame my thoughts, Ruth was guiding me in learning how to control my own attention. She was teaching me to perform the greatest magic trick of all time, an illusion bigger than anything Houdini could pull off, and in front of a really skeptical audience known to heckle at will—my own mind.

By learning to observe my thoughts, I was learning to separate myself from them. At least, that's what Ruth told me. At the time I was not quite sure I understood it all. Still, even with Ruth and her tricks, I couldn't see my life changing all that much. I still lived in a small apartment in a part of town that no one volunteered to live in. I was still poor. I had few friends and a social life that did not exist. And although I knew my parents loved me, my life remained dysfunctional and chaotic. At that time, it seemed that if you were born rich you had it made. If you were born poor, you were like the sucker brought up onto the hypnotist's stage who gets mesmerized into believing he's a bird. No matter how many times he flaps his wings, people are only going to laugh and he's never going to really fly. I tried to open my heart. I tried my best to recite my affirmations. But in my mind I was still the poor kid, living in a small apartment, who was often hungry for food and for love.

I had a story about who I was and what my future held. I wasn't ready yet to see my wounds as gifts. But I was ready for Ruth to teach me her last trick. She had been teaching me every day for five weeks, and we only had a week left before she went back to Ohio.

"Jim," Ruth began, "I know that some of what I've told you, you don't think really has done anything. I want you to know that it has. Far beyond what you can realize at this moment."

I nodded and tried to interrupt her to tell her that it had done a lot, but she didn't let me speak.

"We don't have much time left together, Jim. In the time we have left, I'm going to teach you the greatest magic I know. But you must absolutely listen to everything I tell you. Everything. The reason this is so important is because, unlike everything else we have spent so much time on, this last thing has the power to give you everything you think you want. Unfortunately, because it can give you everything you think you want, it can be dangerous. You need to understand that what you think you want isn't always what's best for you and others. You need to open your heart to learn what you want before you use this magic, otherwise if you don't really know what you want and you get what you think you want, you're going to end up getting what you don't want."

Huh? Say again?

At the time, I hadn't the least understanding of what she was telling me. I only heard "It will get you anything you want."

Finally I was ready. I knew this was going to be the magic trick that changed my life like Ruth promised. I had tried to get her to start on the last trick earlier. I kept telling her my heart was open and let's get on with it and start right away, but she always just shook her head at me.

"Jim," she warned, "you can't skip opening your heart. It's the most important part. Trust me. Promise me you'll always do this first, before this last thing I am going to show you. I know you think of what I teach you as tricks. And perhaps in some ways they are magic tricks. But also please remember such tricks have power. If you don't take what I am saying seriously, there will be a huge price to pay. Learn this from me now, and you won't have to learn it the hard way later."

"I promise." I would have promised Ruth anything to learn her last trick. Open heart or not, it didn't really matter. I already knew exactly what I wanted.

Exactly.

I wish I had listened more carefully. I wish I had learned at twelve to lead with a heart that's wide-open—to others and to the world. What pain could I have prevented? How different would my life lessons have been? What relationships might have worked out that ultimately didn't? Would I have been a better husband? A better father? A better physician? Would I have gone so brashly through the first half of my life demanding my due? What choices would I have made differently? It's hard to say. I believe we learn what we are meant to learn, and some of us are simply meant to learn things the hard way. Ruth tried to help me as best she could. She taught me to stand up for myself and to not let others determine my value, my worth, or my potential. She tried to save me from causing my own suffering. But I was young, and I was hungry, and when she showed me how to train my mind she opened up the whole world to me, and I attacked it like it was the enemy. There's no way I could have known then what I know now, because if I had, I would have truly opened my heart first. The head is powerful, but it can only get us what we really want if we open our heart first.

Experiencing pain can be a gift if one learns from the pain. But when one needlessly causes pain and suffering, not only to oneself but to others, it is neither ennobling nor fair to those who are sharing the path with you. Ruth taught me some very powerful magic, and I could have saved myself, and many others, from a lot of pain and suffering if I had paid more attention to what Ruth was saying that day.

But I was barely a teenager, and paying attention was something I had only just begun to learn.

Ruth's Trick #3

Opening the Heart

1. Relax your body completely (Ruth's Trick #1).
2. Once relaxed, focus on your breathing and try to empty your mind completely of all thoughts.
3. When thoughts arise, guide your attention back to your breath.
4. Continue to breathe in and out, completely emptying your mind.
5. Now think of the person in your life who has given you unconditional love. Unconditional love is not perfect love or love without hurt and pain. It just means that someone loved you selflessly once or for a time. If you can't think of anyone who loved you unconditionally, you can think of someone in your life to whom you have given unconditional love.
6. Sit with the feeling of warmth and contentment that unconditional love brings, while you slowly breathe in and out. Feel the power of unconditional love and how you feel accepted and cared for even with all your flaws and imperfections.

7. Think of someone you care for and, with intent, extend unconditional love to that person. Understand that the gift you are giving him is the same gift that someone gave to you and will make others feel cared for and protected.

8. As you are giving that same unconditional love to one you care for, think again how you feel when you have been given unconditional love and acceptance.

9. Again reflect on how it feels to be cared for, protected, and loved regardless of your flaws and imperfections and think of a person whom you know but have neutral feelings for. Now with intention extend the same unconditional love to her. As you are embracing that person with love, wish her a happy life with as little suffering as possible. Hold that person in your heart and see her future. See her happiness. Let yourself be bathed in that warm feeling.

10. Now think of someone with whom you have had a difficult relationship or for whom you have negative feelings. Understand that oftentimes one's actions are a manifestation of one's pain. See them as yourself. A flawed, imperfect being who at times struggles and makes mistakes. Think of the person in your own life who gave you unconditional love. Reflect on how that

love and acceptance impacted you. Now give that same unconditional love to that person who is difficult or for whom you have negative feelings.

11. See everyone you meet as a flawed imperfect being just like you who has made mistakes, taken wrong turns, and at times has hurt others, yet who is struggling and deserves love. With intention, give others unconditional love. In your mind bathe them with love, warmth, and acceptance. It does not matter what their response is.

What matters is that you have an open heart.
An open heart connects with others,
and that changes everything.

*You can visit intothemagicshop.com to listen to an audio version of this exercise.

FIVE

Three Wishes

My summer was ending with Ruth's promise to teach me the greatest, most powerful, most secret, and life-changing magic trick of all time. I still did not understand what the trick would be, but I imagined that I would become the greatest conjurer the stage had ever seen. Most magicians made doves appear out of a scarf or rabbits from a hat or a fan of cards out of thin air. The trickiest magicians could conjure themselves—magically appearing from out of nowhere onto the middle of the stage. My summer hadn't started out with a whole lot of hope or anything to look forward to, but like a genie who comes out of a bottle and grants three wishes, Ruth was going to tell me how to conjure anything I wanted.

This was the last week Ruth would be here, and it seemed

as if the six weeks had both lasted a lifetime and also gone by in a flash. Six weeks to learn four tricks seemed like a long time, but Ruth told me it often takes people years to learn and master this kind of magic and that I would have to continue to practice and make it a habit over my lifetime. While I came to the magic shop as often as I could, we would continue to practice the tricks each day until I had gotten them. Only then would Ruth agree to move on to the next trick.

I tried not to think about what I would do when she was gone or how I would spend the few remaining days of summer. Thinking about starting school left me feeling anxious. Every time I started to worry, I would practice my breathing and relax my body. Ruth told me that worry was a waste of time, but I still felt worried about school, about my mom, about my dad, about whether we would get evicted come the first of September when the rent was due. Things weren't so great at home. My mom seemed to be getting more and more depressed. My dad had lost his most recent job because he went on a drinking binge and stopped showing up. Now he just sat at home smoking and watching television. He had promised me that the rent would be paid and kept telling me not to worry, but his promises didn't mean much. I was worried. I was worried we would be evicted. I was worried my mom might overdose. I was worried my dad would start drinking and take what little money we had left. And I was worried for my older brother, who would go to the room we shared and cry. I couldn't cry. I was the one who had to keep it together. I was the one who had to track my dad down in the bars and demand whatever money he hadn't spent. I was the one who had to ride in the ambulance when the paramedics came because my mother had attempted suicide again. I was the one who had to protect my brother from the kids who made fun of him.

I walked through the door of the magic shop with the deep sigh of coming home. Neil waved to me from behind the counter. The day before, as I was leaving, he told me about a secret society for magicians. You had to be invited into it, and you had to promise never to reveal your secrets to non-magicians.

"But I will tell you one of the most important secrets," said Neil. "You have to believe in your own magic. This is what makes a magician great. He believes the story he is telling to the audience, he believes in himself. It's not about the illusions, or the applause, or any sleight of hand. It's about the magician's ability to believe in himself and his ability to have the audience believe in him. A trick is never done at the expense of the audience. Magic isn't a hustle or a con. A real magician transports the audience to a world where anything is possible, everything is real, and the unbelievable becomes believable."

I asked Neil why he was telling me this since I certainly wasn't a part of any secret magic society. Yet.

"You're going to do great magic, Jim. I know it. My mom knows it. But you have to know it. You have to really believe. That's the most important thing, and that is the best secret of all magic secrets. Remember that tomorrow when you start to practice your last trick, and remember it even after my mother is gone."

RUTH HAD LIT a large candle and placed it on a small table, more like a TV tray than a table, in the middle of the back office. I had never seen this candle before. It was a tall red glass cylinder with brown and orange swirls around the outside. The candle inside was white and set about a third of the way down inside the glass so the swirls made the flame look like it was moving and dancing. She had the lights off in the room, so it was fairly dim and seemed more mysterious than usual.

"What's that smell?" I asked Ruth.

"Sandalwood," she said. "Good for dreaming."

I wondered if we were going to have a séance or if Ruth was maybe going to bring out a Ouija board. I was excited and nervous like it was my first day all over again.

"Have a seat." Ruth smiled at me and put her hand on my shoulder. She knew I had been waiting for this trick.

She sat down across from me and just stared into my eyes for a few minutes. "Jim, tell me what you want most out of life."

I didn't know what to say. I knew I wanted money. Enough money so that I didn't have to worry about anything ever again. Enough money so I could buy whatever I wanted whenever I wanted it. Enough money so that people would be impressed with my success and would take me seriously. Enough money so that I would be happy and my mom wouldn't be depressed and my dad wouldn't need to drink.

"Be as specific as possible."

I was a little embarrassed to say it out loud, but I did anyway. "I want a lot of money."

Ruth smiled. "How much money? Specifically."

I had never thought about exactly how much money it would take to make all these things come true. I had no idea.

"Enough money," I said.

Ruth let out a little laugh. "Jim, I need you to say out loud exactly how much money is enough money."

I thought about it. I had seen a man drive a silver Porsche Targa by my school often. He must have worked or lived nearby. He looked so cool. I swore one day I would have one just like that. I remembered a classmate whose father owned his own construction company and who had invited me to his house to play. It was huge like a mansion with a large backyard and a gigantic pool and a tennis court. I was going to live in a house like that someday. I remembered my friend's father lying by the pool wearing a gold Rolex watch covered in diamonds that he took off and laid on the table. He saw me looking at it and told me I could hold it. It was so heavy. He told me it was solid gold. I asked him how much it cost, not knowing it was a rude question to ask. He didn't blink an eye and said $6,000. That was a fortune in 1968. I couldn't imagine having that much money to spend on a watch. I told myself that, one day, I would have a watch just like this man had. I remembered later watching *Fantasy Island* and dreaming about owning my own island. I would grant myself wishes. I wanted to have my crooked tooth fixed so people wouldn't make fun of it and I wouldn't be embarrassed by it. I wanted to go to the fancy restaurants, like I saw on television. I wanted to be so rich that places would be named after me. When I had all those things, then I would feel OK. And that's what I wanted most of all— to be OK.

"A lot," I said. "Enough to have everything I want."

Ruth didn't even hesitate after I said this. "How much is enough?" she asked.

I thought about saying $2 million, but I didn't want her to think I was greedy. "One million dollars," I finally said. "That is enough money."

Ruth told me to close my eyes. She had me relax my body. She told me to empty my mind of thoughts. Then she told me to open my heart. I still wasn't sure about the opening the heart business, but I nodded my way through it all. "Now, Jim," she said, "I want you to see yourself having *enough* money. See the million dollars in your mind."

At first I just saw a room full of money. Stacks and stacks of bills from the floor to the ceiling. Ruth asked me what I was picturing in my mind and I told her.

"Jim, I don't want you to see the money. I want you to see yourself as if you have enough money. Do you know what I mean?"

"Not really," I answered.

"There are two ways to picture yourself in your head. One way is as if you were watching a movie of yourself. The other way is as if you are looking out at the world through your own eyes. I want you to imagine what the world looks like to you when you have your million dollars. Try to picture the world through your millionaire eyes. Imagine you already have all the money you want. What do you see exactly?"

I closed my eyes and tried to imagine the future. I saw a Porsche 911 Targa. It was silver. But I couldn't picture anything through my own eyes. I could see myself driving it, but from a distance, like I was watching TV. I saw myself eating in a fancy restaurant. I saw a big mansion, almost like a castle. But when I tried to look at these things as if they were mine, like Ruth said, I couldn't do it. Everything was like a movie I was watching. And even that was hard to imagine for more than a few seconds.

"I thought this would be easy," I said to Ruth, "but it's hard." I told Ruth about the Porsche 911 and seeing myself in it like it was a movie.

"It takes practice and time and more practice. Eventually, you're going to be able to see the Porsche as if you were driving it. I want you to try and think about how your hands feel against the leather of the steering wheel. What does the car smell like? What does it sound like? Look down at the speedometer and tell me how fast you are going. What is the scenery outside? Is it day or night? What does your body feel like to be driving this car?"

"I have to imagine all that?"

"It is a lot of work, but that's the trick. You can have anything you want by visualizing that it's already yours. It's that simple and that hard, all at the same time.

"I imagined myself coming here to Lancaster this summer. I saw myself in this shop, with my son. I could picture how the sun beat against the glass. I saw my hand in Neil's. And I saw a young boy talking to me. I created all this in my mind, and made it real. Long before my trip was planned. I didn't know how I was going to get to Lancaster, but I believed that I would be in Lancaster this summer. In my mind I was already here."

"You saw me?" I asked.

"I saw myself spending time with a young boy. At the time, I thought it would be my grandson. But it didn't turn out that way. It turned out that it was you I needed to spend time with. You see, Jim, I opened my heart before I imagined this trip. I opened my heart and imagined that I would be where I was needed with someone who needed me. Then I trusted it would happen. Things don't always happen the way we think they will, but I've learned that they happen exactly the way they're supposed to happen. I don't know why I was supposed to spend this time with you. But I know there's always a reason. And I know if I'm supposed to spend time with my grandson it will happen. Jim, there is an old saying: 'When the student is ready, the teacher appears.' You were the one who was ready."

I never really learned that much about Ruth's personal life, but forty-five years after this conversation I would learn that Ruth was able to spend the following summer, 1969, with her grandson, Curtis, in Lake Isabella, a little over a hundred miles away from Lancaster. She worked her own magic. And like me, maybe it happened because now he was ready.

Ruth sent me home that day and told me to practice the first three tricks she had taught me, to pay special attention to opening my heart, and then to write a list of everything I wanted to create in my life. "I want you to write a list of ten things you want. Think about what you want to create. Write down who you want to be. And then bring it with you tomorrow."

"I thought I got three wishes, not ten wishes."

"Jim, you can have as many wishes as there are stars in the sky. But we're going to start with the ten you bring with you tomorrow."

Ruth had never given me written homework before, but I did exactly as she said.

1. Don't get evicted.
2. Go on a date with Chris.
3. Go to college.
4. Be a doctor.
5. A million dollars.
6. Rolex.
7. Porsche.
8. Mansion.
9. Island.
10. Success.

I handed Ruth my list the next day. She read through it. "Hmm" was her only response.

"What?" I asked her.

"Jim, did you open your heart before you made this list, like I asked you to?"

I nodded yes. It was the first and only time I ever lied to Ruth, but I wasn't quite sure how to open my heart. I didn't feel like I really understood that part of what Ruth taught me, and I was so anxious to learn how to get anything I wanted that I didn't want to ask her or have to go backward. I only had six more days to learn how to make the things on my list come true.

"I didn't know you wanted to be a doctor."

It was Job Day in the fourth grade—the day when professionals from the community come to talk about what they do for a living. We'd already had a fireman and an accountant and an insurance salesman—none of whom was of much interest to me. The fireman was pretty cool, but he said his job was mostly a lot of waiting around for something bad to happen. The next man was different. He smiled at each one of us. He was a doctor, a pediatrician, someone who only took care of children.

"It's an honor and a privilege to care for people who are sick, especially children. It takes a very special type of person to do this job," he said to the class. "When I was a kid, I had severe asthma and almost died. My mother took me to the doctor, and I'll never forget his smile. As soon as I saw him I knew I wouldn't die and, at that moment, I knew I was going to be a doctor."

He was glowing as he stood in front of the class and talked about his job. "But it's not a job," he said. "It's a calling. A calling that is not for everyone. A calling that requires those who do it to go above and beyond a regular nine-to-five job. You have to work long hours because people are depending on you, and if you fail them it could mean they die." I looked around the room to see if anyone else was as mesmerized as I was. He must have seen me gaping at him, because after his talk ended we went to recess, and he walked up to me and asked me my name.

Although I was a very good reader and did well in some subjects, I wasn't that great a student. I didn't understand the need to study, and while my parents encouraged me I didn't have a place to study or anyone to help me when I needed it. It's hard to focus when a television is blaring or an argument is in progress. My teacher seemed to focus her efforts on the brightest students or those who were always prepared. I can't remember one time when I was asked why I was late or why my homework wasn't done. Usually, the only time I would speak up was to tell jokes that often got me into trouble, and other times I just felt invisible. But for this man, I had a million questions.

"Did you ever see anyone die?" "What about being born?" "Do you give shots?" "What do you do when kids cry in your office?"

I asked him a dozen unrelated questions about life as a pediatrician, and he took the time to answer every single one of them. When it was time for him to leave, he shook my hand like I was an adult.

"Maybe you'll be a doctor one day yourself."

I couldn't imagine going to college or becoming a doctor, it seemed impossible, as far-fetched as my walking on the moon one day, but he didn't seem like he was joking. He looked me directly in the eye and said, "I can tell that you care, and I can tell that you would be a really good doctor. Don't count yourself short." He smiled at me again as he turned and left the room.

"Don't count yourself short" repeated itself in my head. I

wasn't sure what that meant. I didn't count myself short—it was more like I hadn't thought there was anything to count at all.

But in that moment, with no one in my family having ever even gone to college, I decided this was exactly what I was going to do. Become a doctor. I immediately imagined being called over the loudspeaker in the hospital like I had seen so many times watching *Ben Casey* on TV. It's not lost on me now that he was a neurosurgeon. Coincidence? Who knows? But, I will tell you to this day, I can still with complete clarity see him in my mind's eye and hear that loudspeaker.

I told Ruth, "Yes, I want to be a doctor." Then I corrected myself. "I *know* I'm going to be a doctor." I had no idea how to go about making that happen—I had never even dreamed of going to college, much less medical school—but at that moment I knew it would happen.

Ruth clapped her hands together as if I had just done some amazing feat.

"That's it," she said. "That's exactly it."

"What's it?"

"That knowing. You have to know you will be a doctor and then you have to picture it in your head as if you already were a doctor. See the world through your doctor's eyes."

I closed my eyes and tried. It was hard. I could just barely see myself as a doctor, looking down at my white coat. But it was fuzzy. "It's hard to see."

"That's why you have to relax your body and clear your mind of all thoughts first," said Ruth. She walked me through the first exercises again. "Now that I have your attention, it's time to set your intention."

"My what?" I opened my eyes.

"Your intention. If you relax your body, clear your mind, and open your heart—it's easy to set a clear intention. You intend to be a doctor. You are very clear on that."

I closed my eyes again and thought, *I intend to be a doctor. I clearly intend to be a doctor. I am intending to be a doctor, clearly.* I wasn't sure which one was better so I thought them all.

"Now, Jim, imagine in your head you are looking through a window. The window is all fogged up. Like the inside of a car when it's cold outside. Think of your intention as the defrost setting. Set your intention over and over again so that the window gets clearer and clearer. Less and less foggy. On the other side of that window is you as a doctor. The more clearly you can see the image through the window, the more likely the image will come to pass in real life."

I tried over and over again, and eventually, I could see myself in a white coat through the window in my head.

"Keep at it," said Ruth. "Day after day. Week after week. Month after month. Year after year. Whatever you can see clearly through that window in your mind will become real. And the more you can imagine you already have what's in that window, or you already are what's in that window, the quicker it will happen."

"It's really real?" I asked Ruth. "You promise this magic really works?"

"I promise," said Ruth. "I have never lied to you, Jim. And I'm not going to start now. But it takes work, and some things will take longer to happen than others. And sometimes it won't happen exactly the way you expect. But I promise you, everything you put on your list, everything you feel in your heart, everything you think about and imagine with your mind, if you truly believe, if you work very hard, will happen. You have to see it and then you have to go after it. You can't just wait in your room. You actually have to go get good grades, and go to medical school, and learn how to be a doctor. But in some mysterious way you will also be drawing it to you, and you will become what you imagine. If you use your mind and your heart, it will happen. You have my word."

I went home that night and decided I better write down everything Ruth had told me this summer so I didn't forget. I took out my notebook from my box of special things. I turned to a blank page and wrote "Ruth's Magic" across the top. I turned the page and wrote down everything I knew about relaxing my body, calming my mind, opening my heart, and setting my intention. I wrote down everything I could remember Ruth saying, even if I had no idea what it meant. I made notes in the margins and on the sides. I didn't want to forget anything. I copied my list of ten things I wanted into the notebook.

I read the first thing on the list, "Don't get evicted." I read over everything Ruth had said about this last trick. She told me to think of anything I wanted, repeat my intention over and over to myself, and then create the clear picture in my mind. I wasn't supposed to think about what I didn't want. I didn't know how to imagine *not* being evicted.

We had been evicted before. The police had come and given my mom the eviction notice, followed by people hired by the landlord to throw our things on the street. . . . I didn't want to imagine this over and over again, and how do you imagine it not happening when you can see it happening in your head? All our neighbors and my friends watching us get thrown out. No place to go. Being taken to a homeless shelter and all of our belongings taken away to the garbage dump. I didn't want to relive it even one time in my mind. It was too painful.

I thought about what Ruth said, and decided to imagine the opposite. Every day for the rest of the week, whenever I wasn't with Ruth, I spent hours creating a vision of my family being in our home. I saw us paying the rent. I saw us happy. I cleared away the foggy window with my mind.

At times I found myself still imagining the sheriff knocking on the door. It was a horrible knock. Loud and harsh and impossible to ignore. I knew what that knock meant. I also knew that the first of the month was approaching quickly. Ruth would leave, and I would end up without a home. Both images did battle in my mind, but every day I cleared the foggy window more and more and saw my mom paying the rent, and us staying in our apartment. In my head I kept saying over and over, "The rent will be paid. We will not be evicted."

Ruth and I practiced every day that week, up until our very last time together. She would talk me through visualizing myself as a doctor, and I would go home and practice visualizing the rent getting paid. My dad had said he was expecting some money to come in for a job he had done a long time ago, but I didn't believe him. I had heard those stories before. Eviction was looming, but I fought against it with the only power I had—Ruth's magic.

I said good-bye to Ruth on a Saturday morning. She hugged me for a long time.

"I'm proud of you, Jim."

"Thanks, Ruth," I said. "Thanks for everything you taught me."

Saying good-bye was awkward. It seemed like it should have been a bigger deal than it was. Neil had been with a customer and had just sort of waved me off. Ruth was talking about waiting at the store until Neil could close up and take her to the airport. And then that was it. I got on my bike and headed home.

I was in my room when I heard the knock on the front door.

It startled me. I had been thinking about Ruth leaving. Another knock. It sounded angry and insistent. My stomach turned over, and I could feel my heart start beating fast in my chest. I felt stuck to the floor. The knock started up again. I knew my mom was in bed, and my dad and brother weren't home. I had to answer that knocking. There was no one else.

I looked through the kitchen window, expecting to see the deputy sheriff's patrol car in front, and the deputy at the door. Instead there was a man. A man in a suit. I opened the door, and he looked at me, then asked for my father.

"He's not here," I said.

"Please tell your father that I am sorry I couldn't pay him before. Please give him this envelope, and please thank him for his patience."

He handed me the envelope and walked away. I shut the front door and looked down at the envelope in my hand. It had a name and address written on the front of it. I turned it over. It wasn't sealed, so I lifted the envelope flap. I could see money inside. Lots and lots of money.

I ran into the bedroom and gave it to my mother. She opened the envelope and slowly counted the money. There was enough money to not only pay the rent for the next three months, but also to pay some bills and buy food.

I couldn't believe it. The magic had worked. It really worked.

"I have to go!" I yelled at my mom.

I got on my bike and rode as fast as I could back to the magic shop. Ruth was just walking out the door with Neil.

"Ruth! Ruth!" I screamed.

She and Neil both stopped on the sidewalk.

"I'm glad you came back," said Neil. "I meant to give you this earlier." He handed me a bag from the store. "You can still come by, even without my mom here. Anytime."

I said thanks, and he walked over to the car to wait for Ruth.

I looked into her eyes. "It really works," I said. I had tears in my eyes. "The magic. It's real."

She put her arms around me and hugged me while I sat on my bike. "I know, Jim, I know." She stepped away and started toward the car, but then turned back. "You understand it now, don't you? The power you have inside you? You were ready to learn, and I was privileged to teach you. Each of us has that power inside. We each just need to learn how to use it. But remember, the magic I have taught you is powerful. Powerful for good, but in the hands of someone who isn't ready, it can also hurt and cause pain. And also remember, Jim, it is your thoughts that create reality. Others can create your reality only if you don't create it yourself."

I watched her drive off. I thought I understood what she was saying in those last moments we had together, but I didn't understand enough. Not nearly enough. There would come a time later in my life when I truly understood, but before that happened I had to experience what Ruth had meant about such power in the hands of someone who wasn't ready. I was that someone.

I looked into the bag Neil had given me. There was a plastic thumb tip and a few different decks of marked cards.

I thought about Neil for a minute. I closed the bag up. I really liked his magic, but it didn't compare to the magic Ruth taught me. I had something better. Something way more powerful. I was going to get what I wanted. And one thing I knew I didn't want was to be poor or to be looked down upon by people who thought they were better than me because they had money and lived in nice houses and drove nice cars and had good jobs. I was going to have everything. No one was ever going to look down on me. I was going to be a doctor. Someone whom everyone looked up to. I was going to have a million dollars. I would be powerful. Successful. I knew how to do it. Ruth had taught me. This magic was greater than anything I had ever imagined. And all along it was sitting right there inside me. I just didn't know it. I would train my mind. I would practice. I would work harder, do more—whatever was necessary. I knew I had it in me.

We weren't evicted. It was all the proof I needed. Ruth's magic was real, and it was powerful. I crossed that off my list, and I knew I would cross the rest of it off too.

. . .

I HATED LANCASTER. Certainly, my family situation contributed a great deal to how I felt about the place, but if it weren't for Lancaster, I wouldn't have learned the magic that would allow me to accomplish extraordinary things. I am thankful that I was there, at that time, in that place, to meet the right person. The person who changed my brain with her magic.

My reality before Ruth was that I felt lost and that life was an unfair place where some were lucky and some were not. I didn't see any real possibility that I could become someone important or escape the small and miserable world that my parents lived in. After Ruth, I saw the world differently. I saw myself differently. I believed in a world of unlimited possibilities. I could create anything I wanted, and this gave me a sense of power and a sense of purpose. Ultimately, we all have the ability to learn the same magic. I had tapped into the power of my mind, and I was ready to use that power and not let anyone or anything stop me.

Ruth's Trick #4

Clarifying Your Intent

1. Sit in a quiet room and close your eyes.
2. Think of a goal or something you wish to accomplish. It does not matter that the details of the vision are not fully formed. It is important that such a goal or vision is one that does not involve harm to another or bad intent. While this technique could help you accomplish such a goal, it will ultimately result in pain and suffering to yourself and make you unhappy.
3. Relax your body completely (Ruth's Trick #1).
4. Once relaxed, focus on your breathing and try to empty your mind completely of all thoughts.
5. When thoughts arise, guide your attention back to your breath.
6. Continue to breathe in and out, completely emptying your mind.
7. Now think of your goal or wish and see yourself as having accomplished it. Sit with the vision as you slowly breathe in and out.
8. Feel the positive emotions associated with accomplishing your goal or having achieved your wish. Experience

how good it feels to have taken a thought and turned it into reality. Sit with the positive feelings as you see yourself having accomplished your goal.

9. Once you have seen yourself having accomplished the goal and have sat with the positive feelings, begin to add details to the vision. Exactly how do you look? Where are you? How are people responding to you? Add as much detail to the vision as possible.

10. Repeat one to two times daily or more for ten to thirty minutes. Each time begin with the vision of yourself having accomplished your goal. Sit with the feelings. Each time as you look at the vision add more details. It will start fuzzy, but the more times you do the exercise, the more the vision will become clearer.

11. With each time you do the exercise you will find you are refining the vision as your unconscious mind begins having clarity of the intent. You may be surprised what you discover and how you end up achieving your goal. What is important is the goal, not exactly how you get there.

It is with clarity of intent that
vision becomes a reality.

*You can visit intothemagicshop.com to listen to an audio version of this exercise.

PART TWO

The Mysteries
of the Brain

SIX

Apply Yourself

If my life were a made-for-television movie—perhaps one of the ABC Afterschool Specials that began airing in the 1970s—life would have changed dramatically after Ruth's magic kept us from getting evicted. My dad would have stopped drinking, my mother would have left the darkness of depression forever, money would have continued magically to appear on our doorstep, and we all would have lived happily ever after as the perfect, made-for-television, nuclear family. The Brady Bunch would have had nothing on the Doty family.

But Ruth's magic didn't work that way. A genie hadn't been let out of the bottle to grant my every wish in real time. My family was not magically transformed. My dad still drank.

My brother still hid from the world. My mother still battled depression and a seizure disorder. I had been given the magic, yes, but it was up to me to practice it. Perfect it. And continue to believe that the impossible was now possible. I could try to create a new reality for myself, but I could not change the people I loved, no matter how much I might have intended it. They had to choose to change their reality and that did not happen. This is perhaps the most painful part of being a child. Our life is dependent on others and is beyond our control. Often the impact of others' choices can be deeply wounding and leave lasting scars.

I might not have been able to change anyone else's reality, but I knew I could change my own. I knew every single thing on my list would come true, and soon after Ruth left, I had it memorized so thoroughly that I put it away into my special box with my Dale Carnegie book and my magic tricks from Neil. I also had the little notebook that I kept in the box with everything Ruth had taught me written down inside.

I practiced every morning and every night, day after day, week after week, month after month. Just as athletes who visualize or imagine themselves performing a skill over and over again in their head—the perfect jump shot, the hole-in-one, a home run hit high past center field—are changing their physiology and creating neural patterns in their brain that actually enable their muscles to perform in new ways, I was using visual imagery to create new neural pathways in my own brain. The brain doesn't distinguish between an experience that is intensely imagined and an experience that is real. I was training my mind to become a doctor long before I ever applied to college or medical school, simply by visualizing myself as a doctor. Another mystery of the brain is that it will always choose what is familiar over what is unfamiliar. By visualizing my own future success, I was making this success familiar to my brain. Intention is a funny thing, and whatever the brain puts its intention on is what it sees. Have you ever thought about buying a certain type of car and then it was as if you were suddenly seeing that exact type of car everywhere you went? Was it your *intention* that made the car magically appear or was it your brain's focused *attention* that allowed you to finally see what was in front of you all the time? "You get what you expect" can be a simple idea delegated to a New Age, feel-good thought or a powerful example of neuroscience and brain plasticity. Attention is a powerful thing—it can literally change our brains, creating more grey matter in the very areas that help us learn, perform, and make our dreams come true. Ruth taught me to pay attention to what I expected in life. Did I expect to live in poverty? Did I expect my life not to matter because I was on public assistance or grew up in an alcoholic family? Did I expect my worth to not be as great because of where I lived or who my parents were?

Ruth taught me to refocus my attention and intention from my identity as an impoverished child from a neglectful home and move it toward what my mind thought it wanted most. *Money. Rolex. Success. Porsche. Doctor.* These were my new familiars—these were the images I engraved onto the cells and within the synapses of my prefrontal cortex. The prefrontal cortex controls our executive functions—planning, problem solving, judgment, reasoning, memory, decision making. It's what helps us regulate our emotional responses, overcome a bad habit, or make a wise choice. It's the place in our brain that allows us to consider our own mind—what Ruth had already begun to teach me to do. It's also where we learn to feel empathy and connection to others. Ruth taught me the skills to get anything I wanted in life, and I focused my attention entirely on manifesting the future I dreamed about. I had no idea about any of the details that would help me get into college and medical school—in fact I was completely oblivious to the entire process. But intention setting is its own kind of magic, and since that summer in the magic shop, the universe has always seemed to conspire to get me exactly where I needed to be.

Of course, when it came to surviving high school, the universe was nowhere to be found. In retrospect, maybe I should have set my intention more on succeeding at school and focused on one thing at a time, instead of only what life would look like when I was finally someone.

MY HIGH SCHOOL years passed by in a blur. In some areas I did very well, but in others I just passed. I didn't yet have a clear image of what I needed to do to go to college or medical school. I also didn't understand how to ask for help or guidance. Only later did I realize that many people will help if they are just asked. But at that time I still felt I was alone and didn't know how to ask or even what to ask. As a child, not having mentors or people to turn to for advice or guidance has a huge impact on success in life. You can't do it if you don't know what it is. I had wanted to play sports in high school and I had made the cut for the football, basketball, and baseball teams as a freshman, but I soon found out that school sports required both money and parental involvement, and I had neither on a consistent basis. It's hard to be a member of a team when you can't get a ride to practice, or you're unable to show up for a game because you have to stay home and babysit your mother or have to go to a bar on a Friday night and find your father. I liked the feeling of belonging I experienced when on a team— dressed in our uniforms we were all the same, and we shared a common purpose. I never lettered in a sport in high school, although I wanted to desperately, so during my junior year I took out my list of ten things and added this to it: *Letter in a sport in college—get the jacket!*

Knowing I had my list tucked away helped me take life's

disappointments and seeming unfairness in stride, and relaxing my body and calming my mind every evening eased my anxiety about both school and home. I was living for the future that existed in my mind, and it was a far more enjoyable place to live than our small dingy apartment that smelled like mold and cigarette smoke. Unless I was practicing Ruth's magic or sleeping, I tried not to be at home.

It was this desire to be at home as little as possible that made me apply to Law Enforcement Exploring. In order to be a Law Enforcement Explorer Scout you had to be over the age of fifteen, have at least a 2.0 GPA in high school, and be of good moral character. Every Saturday for twelve weeks we went by bus to the Sheriff's Academy in Los Angeles and learned about law enforcement. For eight hours we studied community policing, criminal procedures, self-defense, and gun safety and were drilled in physical fitness. All the Deputy Explorers wore the same khaki shirts and dark green pants. It wasn't exactly like being on a sports team, but I still got to wear a uniform and be a part of something bigger than myself. It was also nice to have somewhere to go on Saturdays. Once we had graduated from the program we were official Explorers and participated in different roles in our local sheriff's department, working side by side with a deputy. One day we might be on patrol, driving around the community and answering calls. Another time we might be in charge of crowd control at various events like parades, high school football games, and the annual Fourth of July fireworks show. Or we could work in the jail alongside the officers who processed and booked anyone who was arrested.

One Saturday night my assignment was to work in the booking area at the sheriff's station in Lancaster. I was helping the jailer and so was given a key. I hung the key proudly on the loop of my pants, and waited for a massive takedown of some criminal masterminds. I imagined the jail filled with prisoners, with me outside the cell, holding the key to their fate. I was powerful with that special key, but for most of the night there was no one around to see me in all my glory.

I filed endless stacks of paper and reports, drank several Cokes from the vending machine, and basically sat around thinking this part of being in law enforcement was pretty boring. Just before my volunteer shift was over, I heard a patrol car pull up outside the booking area and saw a patrolman walk in with a disheveled, handcuffed man. I couldn't see his face. He was obviously inebriated and his speech was slurred. I felt my heart begin to race. This was it. I would soon be putting this criminal behind bars. The patrolman walked past me with the criminal. His shoulders were hunched over, and I still couldn't see his face, but he swayed and stumbled as he walked. I took out my key, knowing that after they fingerprinted and booked him, it would be time to lock him up. The criminal sat down at the desk and that's when he lifted his head up and looked right at me.

It was my father. He looked confused and angry and very, very drunk. I felt my stomach turn over. I quickly turned away from him and went back over to the filing cabinet. I was so ashamed. I had written an entire essay for my Deputy Explorer application about my high moral character. Now what were they going to think of me? I had answered the questions about my family in a very vague way, and had convinced myself that the deputies didn't know how poor I was or that my father was an angry alcoholic who had been to jail many times. Part of the reason I had joined the Explorers was to prove just how different I was from my family.

I opened the filing cabinet drawer and just stared at the rows of files inside. I wished I could use my special key to lock myself away from this place. Why did it always seem that no matter where I went, I couldn't escape who I was and where I was from?

I felt a hand on my shoulder and looked up to see my supervising deputy standing next to me.

"I'm sorry this happened," he said.

I realized then that he must have known who my father was all along. I could feel the heat flush across my face, so I kept my head down. I wasn't going to cry, but I wasn't sure what I was supposed to do. Was I really going to have to lock up my own father?

"I talked to the officer who brought him in. We're not going to press charges. We'll let him sober up and give him a ride home."

I nodded, and murmured, "Thanks."

I wanted to just disappear, but my supervisor was still standing there with his hand on my shoulder.

"Jim," he said quietly.

I looked up and into his eyes, expecting to see judgment or, even worse, pity. But I didn't see either. And in that moment, I remembered Ruth once telling me that just because something is broken doesn't mean everything is broken. I had always assumed that people judged me because of my father, because of my poverty, because of all the things I didn't have— but feeling that deputy's hand on my shoulder, seeing his eyes full of kindness, I realized that this was how I judged myself. I was poor. My father was an alcoholic. But I wasn't broken. Everything didn't have to be broken just because something was broken. I didn't have to be broken.

"Yes, sir?" I said to the deputy.

"Do you want to leave or finish your shift out?"

"I'd like to finish it out." And the minute I said this, I knew it was true. My father had his path, and I had my path.

The deputy looked at me again. "You know, Jim, my father was an alcoholic too. I know how you're feeling." I felt one last squeeze on my shoulder and then the deputy turned away and walked back out the door.

· · ·

WHEN YOU LOOK at adults who grew up in an alcoholic home, you will find two common outcomes—either they grow up to become addicts or alcoholics themselves, a manifestation of their own trauma combined with genetic exposure, or they become overachievers, bent on being different from their family of origin and bent on escape. I was the second type. This was part of the reason I joined Law Enforcement Exploring. I liked the prestige of being a part of a select group with high moral character. I'm not sure if I was trying to convince the world or just myself. As was the case when my father was arrested, I couldn't always keep my two very different worlds from occasionally colliding. Another one of my assignments for the Explorers was to help pack and distribute food baskets for the poor during the Christmas season. We packed big wicker baskets full of tinned pumpkin, white bread for stuffing, sweet potatoes, and of course, a great big turkey. A few days before Christmas, the deputies went around and delivered the baskets.

I wasn't part of the crew that delivered the baskets, but I liked to hear the stories everyone told about what happened when they knocked on people's doors and surprised them with a gift basket. Sometimes people cried, and I had heard one of the officers say, "You'd think they never saw a turkey before."

I felt good when I helped with these baskets. It was a feeling of elation that lasted for days or even weeks. It was the same feeling I got when I practiced quieting my mind the way Ruth had taught me. Ruth's tricks were a part of my daily life. I didn't tell anyone about it, but every morning and every night I would relax my body, calm my mind, and visualize what I wanted in life and who I was going to be. I didn't open my heart. That trick was difficult for me. It was hard to give love to myself because I had somehow internalized that my situation was my fault. It also made me uncomfortable to offer unconditional love and compassion to myself and to others. Especially those who I felt snubbed me or ignored me or treated me badly.

When I saw the patrolman walking up to our front door with a big wicker basket in his arms, I hid behind the curtains and let my mother answer the knock. I was horrified. I'd had the feeling that we were on the list that year. I didn't want to be someone who needed the basket. I watched my mother unpack one of the very baskets I had helped put together earlier that week. The basket was a reminder that we were poor. I didn't want to have to rely on others. Yet without that basket, we would have had no turkey dinner on Christmas. Nobody in my family knew that I had helped pack this gift. It felt good, not because I had packed the basket but because seeing how happy my mother and father were reminded me how much those baskets meant to so many. It is rare to be on both sides of an act of kindness or generosity. On that particular holiday,

I learned the pleasure of giving and the pleasure of receiving. It was a potent collision, and little did I know then how the knowledge of both would inform my adult life.

I STAYED IN THE Deputy Explorer program throughout high school, from the age of fourteen to seventeen. It gave me a sense of purpose and a place to belong, and those two things, combined with my daily practice of Ruth's magic, produced a very subtle alchemy within me. I found that fear, anxiety, and worry were no longer useful emotions to entertain. More and more I could observe my thoughts and emotions without engaging in an emotional response to them. I wasn't sure exactly who I was becoming, but I knew I was no longer the child I had been. My family became just my family, rather than a wound that caused me pain every day. I also had clarity that I was not my father, my mother, my brother, or my sister. I was me. Their actions were not mine. My brother and sister both had their own struggles, and their own fates to follow. My half sister, nine years older, had dropped out of high school, married young, moved away, and struggled to make ends meet. She would die in 2011 from health complications due to a chronic immune disorder and obesity. My brother, who was very bright, had struggled with being gay in a time and place that did not accept that people could love someone of the same sex. He had been bullied often for being different, even though this difference hadn't been named or articulated. He left Lancaster while I was in high school, and for my last two years of high school I felt even more alone. But Lancaster became a place I would someday leave rather than the place I was stuck in. My future wasn't bleak and drab but played out nightly in vivid Technicolor across my mind's eye. I had absolute faith in what Ruth had taught me, and absolute trust that my future was rushing to meet me.

With my senior year already under way, I realized I had to begin thinking about college but I didn't know where to begin. My parents, while encouraging, just assumed that since I said I was going to college it would happen somehow. My guidance counselor didn't even bring it up as an option. His meeting with me was short, informing me that he could give me information about technical schools if I wanted. I hadn't even known there was a guidance counselor until I got a notice that an appointment had been scheduled. While I had done well in some courses, overall my grades were mediocre. I had no real understanding of the necessity of good grades. For me, school had been a place I had to attend, and while I naturally on one hand had wanted to do well, I had no examples in regard to how to study or prepare to succeed in school. I had never had anyone in my family offer to help me with my homework or even tell me I had to do it. While my mother encouraged me to do well, I had no idea exactly what that entailed. I didn't know anyone who had gone to college. I certainly didn't have any money to pay for college. And I had no idea how to apply. Still, I was absolutely, and naively, certain that the following year I was going away to college.

It was shortly after my meeting with the guidance counselor that I tried to think of whom I could ask about how to apply to college. I was sitting in science class waiting for a lecture to begin on the three laws of thermodynamics when I noticed the pretty girl next to me filling out a bunch of forms.

"What are you doing?" I asked. "What's all that?" I wondered if we had some sort of science test that I had missed somehow.

She looked up from her paperwork. "I'm filling out my college application."

I nodded, as if I knew exactly what that entailed. "Where are you going?" I tilted my head sideways, but couldn't see the name of any school on her forms.

"UC Irvine," she said.

"Really?" I wasn't sure exactly where Irvine was, but knew it was south of Los Angeles somewhere.

She laughed a little. "Well, it's where I hope I'm going. The deadline is next Friday for all of this. I'm never going to get it done." She waved her hands over the papers.

I said nothing as my mind shifted into overdrive. Deadlines? I had no idea there were application deadlines. I didn't know how any of this worked, and for a moment I felt doubt creep in. Would I even be able to apply to college in time?

"Where are you going?" she asked.

I thought for a second, considering how to reply. "I'm going to UC Irvine too." I don't know why that came out of my mouth, but in that instant that was my first-choice school. I didn't really know anything about UC Irvine, but it was still more than I knew about any other college. I knew I needed to go to college to become a doctor, but no one had ever told me there were deadlines and stacks of forms to complete.

She looked at me and said, "I guess you've already filled out your application?"

I stared at her for a moment and then lied. "Well, no . . . I haven't received the application. I thought it was due next month. I've been waiting for it."

Then, like a magician, she pulled out another set of forms and said, "Hey, you're in luck—I have an extra application. Do you want it?"

"Sure. Thanks." I took it from her. I went home that night and tried to fill it out. I realized I needed to get my transcripts, letters of recommendation, and my parents' tax return. For the next three days I ran around getting everything. I filled out the forms to apply for financial aid and hoped it would be enough to pay for school. It was during this time that I really looked at my grades and test scores and the average grades and test scores of those who were accepted. I would never get in. What had I been thinking? I realized all of Ruth's magic wasn't going to help. Plus, I didn't have the money for the application fee. I mailed the application anyway. When I got home, I sat on my bed and thought about Ruth. All the stuff she taught me. Could it really work? That night and every day thereafter, I sat on my bed and visualized receiving my acceptance letter. UC Irvine was the only college I applied to, and for a few months I heard nothing. In that time we had moved twice. When the thick letter from UC Irvine finally arrived it had multiple forwarding notices on the outside. I took it up to my room and sat down on my bed. I slowly breathed in and out, in and out. I knew Ruth had been right.

I had applied myself to my "practice" every day for years, and I had applied to college. I stared at the large white envelope, and I saw myself in a white coat someday. This was the next step in the universe's conspiracy plot to make me a doctor, and as I ripped open the letter I had no doubt what it would say.

Congratulations on your acceptance to the University of California at Irvine. . . .

My future had arrived. Yes, it had to be forwarded many times through the mail, traveling from one seedy apartment to the next, but my future had chased me down and finally found me.

"Thank you, Ruth," I murmured. "And good-bye, Lancaster."

I had been accepted. Amazingly, by graduation I had significantly improved my academic performance and had received some small scholarships and enough financial assistance to pay for tuition, room, and board. I was going to college.

I was free.

I STILL VISUALIZE what I want in life. I see it through a window in my mind that often isn't quite clear and then I trust with absolute faith that when the time is right, it will be crystal clear. I have learned that this process of manifesting isn't always linear and doesn't always operate on a timeline that is as I desire or makes sense, but whatever I visualize usually manifests, and when it doesn't there clearly has been a good reason it didn't. Over the decades I have learned that having faith in the outcome is quite different from being attached to the outcome, and I learned the hard way that you have to be careful about what it is exactly that you want to manifest. I have also learned that there is immense power contained in one's intent.

I HAVE NEVER BELIEVED in a powerful Supreme Being who decides who is worthy and who is not, and grants wishes and gifts accordingly. I have seen too many times the arbitrariness of a world in which an incredibly kind and wonderful person meets a sudden and painful death, and I have also seen people who are fundamentally unkind and even evil flourish. But I do believe we have the ability to transform the energy contained within each of us to have a profound impact. Each of us can change our brain, our perceptions, our responses, and even our fate. This was what I learned from Ruth's magic. We can use the energy of our minds and the energy of our hearts to create anything we want. It still takes hard work. It still takes consistent effort and intention. I didn't take a magic pill and suddenly become a neurosurgeon. But I learned as a teenager that I had the choice of how to use my mind and how to respond to events around me and, later in life, how to use my heart to touch those around me. I don't think there is a law of physics that can adequately describe the power and force that's created when you use both, but I will always remember the first law of thermodynamics that we had to memorize in science class the day I was given a college application.

Energy can be neither created nor destroyed. However, energy can change forms, and energy can flow from one place to another. That is the gift we are each given.

The energy of the universe is within us. It is in that stardust that makes up each of us. All that power of creation. All that power of expansion. All that beautiful, simple, synchronized power. Energy can flow from one place to another. And it can flow from one person to another. Ruth taught me my first lessons, and life has taught me the subsequent lessons. I have spent many years proving the reality of what I learned in the magic shop, but ultimately it comes down to one simple, mysterious fact. We can study every single mystery of the brain, but its greatest mystery is its ability to transform and change.

There are times when I wish I had a scan of my brain at twelve, and then again at eighteen, and yet again after every hard truth that my brain has had to grasp over a lifetime. I was off to college with a changed brain, and studies have proven that focused meditation like Ruth taught me increases the ability to concentrate, to memorize, to study complex ideas. Would I have gone to college and medical school if I had never met Ruth? Probably not. Would I have succeeded in both if I hadn't unknowingly prepared my brain for the academic rigors that the next twelve years would bring? Most definitely not.

When our brain changes, we change. That is a truth proven by science. But an even greater truth is that when our heart changes, everything changes. And that change is not only in how we see the world but in how the world sees us. And in how the world responds to us.

SEVEN

Unacceptable

Just under the cerebrum, and in front of the cerebellum, sits the brainstem. If you imagine the cerebrum as a world-famous rock star on a concert tour, the cerebellum would be the choreographer, determining the moves the cerebrum makes, and the brainstem would be the road manager—responsible for coordinating all the information needed to make sure the tour runs smoothly and the rock star has everything he or she needs to be a rock star. The brainstem is much smaller than the cerebrum, but it is in charge of all the functions that keep the body alive, and it is the highway that is responsible for millions of messages that need to pass back and forth between the brain and the body.

The brain begins forming approximately three weeks after

conception when the neural tube fuses shut and the first synapses of the central nervous system allow for fetal movement. The brainstem then develops and coordinates the necessary vital functions such as heart rate, breathing, and blood pressure—creating the potential for life outside the womb. The higher regions of the brain—the limbic system and cerebral cortex—are primitive at birth, allowing time for experience and the environment to shape them completely. This shaping and development of the higher regions of the brain through experience never ends—there is no retirement for the brain—every experience matters.

Noel came into the emergency room complaining of headaches, nausea, and vomiting. She had her husband and two children, a four-year-old girl and a six-year-old boy, in tow. The couple were in their early thirties, and Noel was eight months pregnant. Headaches and nausea can be normal symptoms of pregnancy, but in the third trimester, their sudden onset along with high blood pressure can be an indicator of preeclampsia, a dangerous condition for both mother and baby. I happened to be on call that morning, making rounds in the hospital, when the family came in. The obstetrician had been called but had yet to arrive at the hospital when Noel suddenly collapsed in the emergency room and became unresponsive.

By the time I got to her, she'd been intubated and was undergoing a CT scan of her brain. During the scan her vital signs started going crazy, and her blood pressure became incredibly unstable. Looking at the scan, I could see that what was once her brainstem had now been almost completely replaced with blood. Noel had sustained a massive brainstem hemorrhage—an intraparenchymal bleed—the kind people don't recover from. We began resuscitation efforts right there in the CT scan suite, but I held out little hope. I saw no sign of brainstem reflexes—those involuntary movements that occur when the brainstem is functioning properly. Her pupils were fixed and dilated. She was completely unresponsive.

Noel's body was still alive, but her brain was dead.

I ordered medications to sustain her blood pressure, and called the operating room to tell them to get ready.

"Page an OB immediately," I yelled at the nurses. "This baby needs to be delivered now, or it will die."

I ran alongside the gurney, heading to the operating room, praying that an obstetrician would show up. The OR team had rapidly set up for an emergency C-section. We wheeled her into the OR. The pediatrician was there, but no obstetrician. Noel's blood pressure began dropping rapidly, and her heartbeat was becoming more erratic. And suddenly everyone was looking at me. Time was running out. It had been twenty years since I had rotated on obstetrics as an intern, but there was no other surgeon in the operating room. Unless I did something, this baby was going to die. I was going to have to perform an emergency cesarean section and deliver the baby.

There was no time for preliminaries or any more hesitation. Noel was brain-dead. I knew we wouldn't be able to sustain her blood pressure much longer.

We placed her on the OR table. The anesthesiologist quickly anesthetized her, and I rapidly prepped and draped her for surgery. I looked around again praying for the obstetrician to walk in. Her heart suddenly began skipping beats with the blips from the electrocardiogram (EKG) machine. The anesthesiologist looked at me and said, "Her pressure is dropping. We've maxed out on the drugs. You need to move." I could feel the sweat on my forehead and realized I was breathing fast. I was scared. And then I closed my eyes and began breathing slowly. In and out, in and out. I was back at the magic shop. I took a scalpel and sliced open her abdomen and then her uterus. I placed my hands into her body and pulled out the baby. There was a small thin cut across the baby's forehead from the knife I had used to open up Noel, but apart from that, he was alive and healthy. I handed him to the pediatrician, cut and clamped the umbilical cord, and sewed Noel back up.

Her heart stopped beating just seconds after her baby boy was born.

They don't give you any training in medical school on how to tell a husband and two young children their wife and mother is gone. You can't be human and not feel the pain of the relatives. Wave after wave of grief, anger, denial, and despair.

That is why so many doctors will simply say, "I did all I could. I'm sorry." Then they will immediately walk away, leaving a hospital chaplain or other staffer to pick up the broken pieces. There is nothing matter-of-fact in telling a husband his wife has died. No *sorry* that can ease the pain of a child who can't begin to fathom that this one horrible day means his mother will never make him a peanut butter sandwich again, or read him a story, or kiss and cuddle him after he's fallen down.

I took Noel's husband aside and told him what happened. He closed his eyes, reached out to me, and wailed a horrible cry of pain and despair. There was nothing I could do but hold him as he cried. The two children, seeing their father cry, also began to wail. I did my best to make space for this family's grief. I tried to tell Noel's husband about the baby, but he couldn't hear anything beyond the hard truth that his wife was gone.

As I sat there with them I noticed that the front of my surgical scrubs was splattered with tiny drops of blood. Noel's blood? Blood from the baby's forehead? Did it matter? It's hard to celebrate a birth when you are grieving a death, but isn't that what it all comes down to in this life? We are born and we die, and everything that happens between the two can feel so random it defies logic. The only choice we have is in how we respond in each precious moment we are given. In that moment, there was nothing but pain, and my choice was whether to offer comfort and share the pain or to walk away.

I stayed with them, but for how long I don't know. I just know I was there for them as best as I could be.

Noel's brain had died and all those functions each of us take for granted ceased. And here was her son whose brain was now experiencing the reality of the world for the first time. Again the randomness and arbitrariness of the world. Our experiences and our environment shape us all, and my hope was that this family would recover from this tragedy, and this baby would not carry invisible wounds from the story of his birth and the randomness of his mother's death.

It wasn't my first death as a surgeon, nor would it be my last. It also wasn't the first time I had walked away from a family with blood on my clothing.

The first time that happened I was going off to college, and the family was my own.

THE NEWS I had been accepted to UC Irvine was met with both excitement and disbelief by my parents. I had talked about going to college, but I don't think they had connected my desire with the reality of me getting accepted and leaving. As the date of my departure approached my father disappeared. Whenever there was stress or a significant event was about to occur, my father couldn't handle it and left to have his fear and his anxiety lessened by his drug of choice, whiskey. The night before I left to go off to college, I paced around our tiny apartment with excitement and nervousness. Everything I owned could fit into one large duffel bag, and I was all packed by bedtime and ready to make my escape the next day. I was even sleeping in the clothes I was going to wear during the ride to Irvine just so I wouldn't have to pack anything else in the morning or leave anything behind. I wasn't sentimental or nostalgic. I was just ready to leave. My father had now been gone for almost a week, and while he knew the date I was to get on the bus to Irvine, I wasn't sure if I would see him before I left.

I told myself I didn't care. But I did. I loved my father with all his failings. When he was sober and present, he was funny and smart and kind. He was my dad.

It was around three in the morning when I heard the yelling and pounding and then more yelling. My father was at the front door, extremely drunk from the sound of things, and locked out.

My mother stumbled out of her room in her bathrobe, and I saw the terror on her face. She didn't make a move to open the door, and I could see her staring wide-eyed at the front door. She put her hands up to cover her ears, and I could see that she was shaking and trembling. We debated calling the police.

The yelling outside the door grew louder, and I knew it wouldn't be too long before someone else called the police. I had a bus to catch in a few hours and I didn't want to miss it because I had to spend the rest of the night dealing with the police as they arrested my father. I took a step toward the door just as my father kicked his foot through the cheap plywood, splintering the door almost in half. I saw his arm reach in and turn the knob.

He stepped through, yelling even more loudly than before.

"Goddamn it—don't you ever lock me out of my own house again!" he screamed, looking straight at me. His face was contorted and his eyes were dark and wild. My mother began to move to the corner of the room, and this caught his attention.

"Why the hell didn't *you* open the door?" He began moving toward her, and she quickly started backing up until she was up against the wall. I had never seen my father this angry. Usually when he drank he would just eventually pass out. He had never been physically violent.

"Don't go any closer," I heard myself say. I wasn't sure if he heard me or not, and he took another step toward my mom, who looked like a little fluttering bird inside her oversize bathrobe. I had never stood up to him before. We had all been complicit in accepting his behavior and his drinking. But it wasn't acceptable any longer. Not this time.

I stepped between them and yelled louder to get his attention. *"If you move one step closer, I'm going to have to hit you. I will, I really will."*

He ignored me and took another step toward my mother. It felt like I was moving in slow motion or trying to move underwater as I stepped forward and raised my arm. I made a fist and aimed for his nose. I heard and felt the bone crack. Then he fell, hard, like a tree.

My mother screamed, and I watched him land on his face as blood spurted and splattered everywhere. I could smell alcohol mixed with a tangy, coppery, metallic smell that I knew was blood.

Lots and lots of blood.

The bile rose up in my throat and the nausea was overwhelming. I lurched my way to the bathroom, barely making it before the vomit came. I kneeled in front of the toilet and murmured the closest thing to a prayer I have ever said. *Help me.* I wiped my mouth with my sleeve and returned to the living room. My father was still facedown, not moving. Had I killed him? I turned him over. Blood and snot streaked his face. I had never seen so much blood. His nose was off-kilter and twisted awkwardly to the left side of his face. *What a mess*, I kept thinking. *What a terrible mess.*

I heard him moan a little, and as he regained consciousness, I put his head in my lap. I didn't even realize I was crying until I saw a tear land in a small pool of coagulating blood on his cheek. The punch had sobered him. He slowly looked up, and he studied me in a way I'd never seen before. He said, "It's OK, son. It's OK." My mother continued to cry, but I wiped my eyes dry. In that moment, I knew that everything would be different between my father and me.

It was now 6 A.M. and my bus was leaving at 7:30. My mother was attending to my dad who was now pretty sober and sitting in the chair drinking coffee with cotton balls stuffed in his nose. He looked at me again and then looked down. My mom told me she didn't want me to miss the bus. And in that bizarre moment I kissed them both, gave them a hug, and walked through the splintered front door and left home for college. As I was walking toward my friend's car, my ride to the bus station, I noticed some blood spattered on the front of my pants. It was too late to go back and change. And anyway all my clothes were in the duffel bag. I wasn't sure what the good-byes were like for other kids going off to college for the first time, but I was pretty sure it wasn't anything like this.

EVEN THOUGH I had been accepted into college, I was ill prepared for juggling my full-time job with classes and studying. I also rowed crew—determined to get my letterman's jacket. Year after year, it seemed like I studied harder than anyone else—just to get a passing grade. I rode the bus from Irvine to Lancaster often during the first few years of school, and other times I hitchhiked. Even though I worked hard, the weeks I left school to take care of my mother, manage my father, or help them dig out of one crisis or another added up. When the time came to apply to medical school, not only did

I have a GPA of 2.5, it looked like I wouldn't even graduate. As a premed student, I was failing miserably. The average GPA for acceptance to medical school at that time was almost 3.8.

I still felt deep down that I would become a doctor. The image of me in a white coat wasn't imaginary; it felt as real to me as if I were looking at myself in a mirror. For almost seven years I had mapped that image onto my brain, and not making it a reality was unacceptable to me. And even though this was a reality in my mind, I found that several of my fellow students were happy to remind me that with my grades I would never get into medical school. Unfortunately, so many people allow others to decide what they can or cannot do. This was another gift that Ruth gave me—the ability to believe in myself and accept that not everyone will want me to succeed or accomplish great things. And how to be OK with that reality and not react to it.

The process of applying to medical school began at the end of my junior year. I discovered that part of the application process for UCI students involved obtaining a letter of recommendation following an interview from the premed committee. I dutifully went to see the premed committee secretary to schedule my interview.

I can still see her clearly in my mind, now over a quarter century later, as she pulled out my file and perused it briefly, then looked up at me in a dismissive fashion and went back to flipping through pages. Finally she closed the file and said, "I'm not scheduling an interview for you. You'll never get into med school. It's just a waste of everyone's time."

I stood there dumbfounded. Getting a letter from this committee was imperative. It was the first step in a long list of steps I needed to take in order to apply to medical school. After that there would be application forms to fill out, essays to write, and then hopefully an invitation to interview at a medical school. There were hoops to jump through, and all I wanted was a chance to jump through them.

I took a deep breath. "I appreciate what you said, but I want an appointment."

"I can't do that. You don't qualify." She tapped her finger up and down on the file.

I knew I was so much more than whatever was in that file. That file wasn't me. That file didn't show that I worked twenty-five hours a week while carrying a full load. It didn't show how many times I had left school to deal with complex family issues. It didn't show me getting up every morning at 5 A.M. to row. It really only showed one thing—my GPA—and if that was the only criterion for receiving a letter of recommendation, then the secretary was right. I would never get into medical school. But that file wasn't me.

Ruth had taught me as much, and my continued practice had helped me discover it for myself. She had also told me that I never had to accept the unacceptable. I had to fight for myself. I had overcome too many obstacles, and there was no way this committee was going to stop me. I had to talk to them.

"That's unacceptable."

"Excuse me?"

"I'm not leaving here until I have a meeting with the committee scheduled." I said this calmly and quietly, and I stared directly into her eyes.

"I really . . . can't do that," she repeated.

But I had heard a slight hesitation in her words, a gap that gave me hope. "Look," I said, "I know I don't qualify. I know you usually don't do this. But you can do this. I just need a chance."

She shook her head again.

"I'm not trying to waste your time or the committee's time, and I'm not trying to be difficult. It's just that I'm really not leaving here until I have a meeting scheduled. I don't care how long I have to wait. I just can't accept that I'm a lost cause. I won't accept it."

There was no anger in my voice, and I think she must have heard the absolute conviction and truth in my words. She stared into my eyes for almost a minute.

"OK," she finally agreed. "Next Tuesday, three o'clock."

"Thank you. I really appreciate it."

As I turned to leave the office, I heard her mutter her final words on the subject. "This is going to be interesting."

On the day of the meeting, the dean of the School of Biological Sciences took the place of one of the regular committee members. Apparently he was intrigued, and my audacity in demanding an appointment had spread throughout the committee.

The secretary greeted me solemnly and opened the door into the conference room. A long, rectangular table was at the far end of the room, and the three professors including the dean sat stone-faced, arms folded, at one end. Not a single smile. Each had a copy of my file and transcripts in front of them. There was a single folding chair for me at the other end. Three to one . . . it didn't seem fair. I was twenty years old.

I walked in, looked around, and realized this wasn't a meeting. It was an inquisition.

And I was the heretic.

"Mr. Doty," began one committee member, a chemistry professor whose class I had barely passed the previous semester. "You have several incompletes in classes, and your academic record does not indicate that you will even graduate much less be a successful candidate for medical school. It does not indicate that you will be a successful medical student or hold any assurance that you have the discipline or intelligence to be a physician."

"I believe this meeting is really a waste of time for everyone here. Can you convince us differently, Mr. Doty?" said another member of the committee, a female professor known to be very tough, although I had never met her before. "I appreciate that you forced the secretary to schedule this appointment, but expecting us to recommend you for a profession for which you have zero chance of success is the height of arrogance. Medical school is extremely competitive, which I'm sure you are aware, while your GPA is not."

I looked to the dean of the school. But he said nothing, just stared at me curiously. He was only there to observe.

"I would like to say something," I said.

"We have other meetings scheduled, and you are free to make your case, but make it brief."

The folding chair I sat in was small, and reminded me of the chair I had sat in for hours across from Ruth in the magic shop. Ruth taught me not to let circumstance define me. Not to let other people define my worth. Yes, there was no doubt that my grades were terrible, but there was more to it than that. I took a deep breath and stood up.

"Who gave you the right to destroy people's dreams?" I paused for a moment and then continued. "When I was in fourth grade I met a man, a physician. He planted a seed in me that someday I could become a doctor too. It didn't look likely. No one in my family had ever gone to college. No one had ever been a professional of any kind, much less a medical professional. In eighth grade I met a woman who taught me that anything is possible if you believe in yourself, if you stop the voice in your head that tells you who you are is based on who you were. I grew up poor. I grew up alone. My parents did the best they could, but they had their own struggles."

I looked at the committee members. The two professors still had their arms crossed, but the dean had leaned forward a bit. He gave me a slight nod to continue.

"I have had this dream for most of my life. It has driven me. Sustained me. Been the only consistent in my life. Yes, I haven't always had the best grades, but not everything has been in my control. I have worked as hard or harder than most, and even if my record doesn't show it, I will guarantee you that there is no one who has come before this committee more determined than I am to succeed in medical school."

I looked at these three who held my future in their hands. Two of them didn't seem to be listening, and for the first time in a long time I felt fear and anxiety course through my body. I knew this feeling. It was what the first twelve years of my life had felt like. My heart began racing. I felt like that lost boy all over again, and the doubt began to drift through me like mist in a fog. Who was I to think I could become a doctor? These were the people who knew best. And then suddenly I could hear Ruth's voice in my head telling me to open my heart. I closed my eyes, and I saw Ruth's smile. *You can do it, Jim*, she said. *You can do anything. You have the magic inside you. Let it out.*

I continued to pour out my heart for what felt like forever. I told them about growing up poor and my struggle to get into college. I told them about my mother and my father. I told them about the many times I had to leave school to take care of my parents. I told them about how hard I worked in school just to maintain my grades and stay enrolled. It was amazing that I was even standing before them wanting to go to medical school, and I did everything I could to make them see just how extraordinary it was. "You know there is not one shred of evidence that a high GPA correlates with being a good doctor. A high GPA doesn't make you care. Every single person, at one time in life or another, needs a chance to do something no one else believes is possible. Each of you here today is here because someone believed in you. Because someone cared. I am asking you to *believe* in me. That's all I'm asking. I am asking you to give me the chance to become who I dream of becoming."

There was silence for a moment when I was done. They told me they would consider all that I had said.

The dean then stood up and shook my hand. "Jim, I think you have given us a perspective that too often we ignore. We forget it's a human being who sits before us, not a file. While many have fulfilled all the criteria we require, in many ways, the criteria are arbitrary. It took nerve to come before us. It took passion and bravery to share what you shared. You don't give up, do you?"

"No, sir," I replied. "I do not give up. Thank you for your time," I said as I left the room.

The secretary looked up at me when I walked by.

"How'd you do?"

I shrugged my shoulders. Only time would tell.

She smiled at me warmly. "I overheard a little of that in there. I have a feeling everything is going to work out for you." She handed me a flyer. "You might want to take a look at this. The deadline has passed, but my sense is that deadlines for you aren't acceptable either."

The flyer was for a summer program called MEdREP at Tulane Medical School. It was a program for minority and economically disadvantaged students hoping to pursue a career in medicine. It was a summer enrichment course that gave you lab experience, and helped you prepare for the MCAT, the test every medical school applicant is required to take.

"Thanks," I said. I stared down at the flyer. Tulane Medical School. I didn't know anything about Tulane, but in that moment, I had a feeling it would be key to my future.

The premed committee ended up providing me with the highest recommendation possible. Ruth's magic had worked again.

And when I called the summer program, the person who answered informed me that the deadline had passed. I asked to speak to Dr. Epps, the program's director. I told her I had to be accepted into the program. She let me tell my story and finally she said, "Jim, send in your application. It will be fine." And two weeks later I had the letter of acceptance to the

MEdREP program in hand. Unfortunately, I didn't have the money for airfare to get to Tulane, which is in New Orleans. Coincidentally, right after I got the acceptance letter to the program, I received a call from my father. He was in jail in Los Angeles and was about to be released and needed me to come get him. He said he needed money for food and a hotel room because my mother wouldn't let him back in the house, and he would end up sleeping on the streets. I only had money for my food and my rent that was due in two weeks. He told me he was expecting a check shortly. *Here we go again*, I thought. But I knew I would help him. He was my father. My friend Keith, who knew some of my family's history, offered to drive me to Los Angeles to pick up my dad. Dad actually seemed OK, as he had been in jail for several weeks and sober during that time. We took him to skid row and rented him a room for two weeks and I gave him $200. I told him about the summer program in Tulane, and he smiled and said he was proud of me. He thanked me. I still had no clue how I was going to pay to get to Tulane, but two weeks later an envelope arrived with writing I recognized as my dad's and in it he had signed over a check to me for $1,000. My father had given me his last cent so I could go to New Orleans. I cried. That summer program was transformative. It exposed me to lab research and allowed me to meet several medical school faculty members. It prepped me to take the MCAT, and it gave me experience being interviewed. It was an intense summer of work, but I was completely focused and completely happy. I was going to be a doctor. I was sure of it.

In the fall I applied to Tulane and waited anxiously. I knew that I had done well during the MEdREP program, and I had done well on the MCATs, but because of my GPA I knew my application was not competitive compared to the vast majority of applicants. I was also working two jobs and the long hours were taking their toll. It was hard to stay focused. It was during this time that I got a call from my mother. My dad had been drinking heavily and had decided suddenly to leave on a Greyhound bus and visit family in Kentucky. She was worried as he had taken nothing with him, and it had been two weeks with no word from him and he hadn't shown up in Kentucky. While my dad would disappear at times, I couldn't recall a time when he was gone that long without reappearing or us hearing from him or from a jail. I now added this to my worry list. My mother called back a few days later saying that my father was in the veteran's hospital in Johnson City, Tennessee.

IT WAS EVENING but I immediately contacted the hospital and spoke to the doctor on call. My dad was in the intensive care unit receiving high-dose antibiotics and on a respirator. He was only intermittently responding to commands. He had severe pneumonia and they were having difficulty oxygenating his lungs. The doctor indicated that Dad seemed to be responding but it was still touch and go. He asked me for a bit more background and his medical history and I realized I knew very little about my father. I didn't know if he had any ongoing health issues. I didn't know if he had been on medication, had ever been operated on, if he had allergies . . . all I knew was that he drank. My entire knowledge of my father related to his drinking.

As I hung up the phone, I tried to think of times he and I just sat and talked or just did something together. Something not related to his drinking. There were only vague, out-of-focus memories. Nothing I could hold on to. Now he had gone off on a bus to see relatives and had never made it to them. What had he done on that bus? What had he been looking for? Why had he chosen to go so far away at this particular time? They were futile questions, and I knew ultimately it was his drinking that had led to his being in a far-off hospital all alone.

I sat down on the side of my bed and cried. I needed to get there, but I had no money. My mother had no money. I had exams coming up. The next few days were spent worrying. I called the hospital several times. He was no longer conscious and his organs were failing. The doctor told me the prognosis was poor, and he was probably going to die. My roommate offered to lend me the money for the plane ticket. I made arrangements and planned on leaving at noon the following day. I had no idea what I would do once I got there. I just didn't want him to be alone.

I fell asleep but I was restless. I had never been on an air-
plane. I didn't know anything about the place I was going. I
was scared. I was tired. Finally I did fall asleep, a deep sleep.
Suddenly I woke up. I wasn't sure what woke me up. I was just
up and awake with my eyes wide-open. I looked around, and
at the end of my bed was my father. He looked at me. He
looked well. Better, in fact, than I had seen him for a long
time. He was calm and had a look on his face that wasn't a
smile but a look of kindness and acceptance. He said, "Hello,
son. I came to say good-bye. I'm sorry I wasn't the father I
wanted to be. I'm sorry I wasn't there for you. Each of us has a
path. I had to take mine. I want you to know that I'm proud of
you and love you very much. I have to go now. Remember that
I love you. Good-bye, son." I said, "I love you too, Dad." And
then he was gone.

I sat up. I wasn't sure if I was dreaming or if it was real. I
didn't know what to think. I just sat there thinking that when
I saw him I was going to hug him and tell him it was OK.
That I loved him. I fell back asleep until the phone rang and
woke me up. I picked up the phone slowly, half-awake. It was
my father's doctor. He wanted to tell me that my father had
passed away an hour before and that he was sorry. He said that
right before he died, he opened his eyes and smiled. He wanted
to let me know that he wasn't in any pain when he died. I
thanked him and hung up the phone. I called my mom and
we both cried. She said he had done everything he could and

that deep down he was a good man and that he loved me very much.

My father did love me.

I knew he loved me.

And I loved him.

LESS THAN A YEAR after I went before the committee at UC Irvine and two weeks after my father passed away, I was accepted into Tulane University School of Medicine. When I received the acceptance letter I went into my bedroom and sat on the side of the bed and slowly opened the envelope and thought of my father. I looked over to where he had been that night when he visited me and said good-bye. I knew he was proud of me.

As was pointed out during my interview with the premed committee, I didn't have enough units to graduate, but I still walked in the graduation ceremony along with the rest of the graduating class of 1977. My acceptance to medical school was conditional on receiving my diploma. My junior year, I had returned home to take care of my mother after a suicide attempt and had to drop all of my classes. As a result, I was short three biology electives. There was no way for me to complete them before medical school began in the fall. I had overcome so much and now it was all at risk. I didn't know what to do and then I realized all I could do was to tell the truth. I picked up the phone and called Tulane and asked to speak to the dean of admissions at the medical school. I waited for what seemed like eternity and he came on the line. He seemed to know exactly who I was. I explained the situation and there was silence. Another eternity. He said, "Jim, we want you at Tulane. If Irvine allows you to transfer credits from medical school to fulfill your missing electives then you're set." I must have said thank you a million times and hung up. It was amazing what happened next. I explained to the professors whose classes I had dropped that I was accepted to medical school but because of a family emergency I had to drop my courses during the last quarter and would they consider allowing me to transfer a medical course to fulfill the requirement. Each one immediately agreed, congratulating me on my acceptance. I didn't realize till later that they had all assumed that I had a stellar GPA and MCAT scores and, of course, they would overlook not completing the elective and substitute in a course from medical school.

Sometimes rules and criteria are critically important, but oftentimes they are arbitrary and act only to sift through numbers and limit opportunity. Having straight A's or an undergraduate degree is an arbitrary barrier to becoming a doctor. I knew I had the innate intelligence and the determination to be an excellent physician.

Now was the time to prove it.

EIGHT

It's Not Brain Surgery

I never planned on becoming a neurosurgeon. My plan was to become a plastic surgeon—I had been moved by children with craniofacial disorders and was attracted to the technical complexity of the surgery. Seeing pictures of children with facial deformities struck a nerve in me. I had a special empathy for those children who had wounds they couldn't hide from the world and who had to constantly see others turn away from their disfigurement. But also, I really liked aesthetic plastic surgery and imagined being a university professor caring for children part of the time and then going to my Beverly Hills office to see my rich private plastic surgery clientele. Plus, being a plastic surgeon to the rich and famous paid very well, and I would meet a lot of very attractive women.

I had accepted a scholarship to pay for medical school for the first year, and after my freshman year I accepted a scholarship from the army. I felt a deep obligation to serve my country, and I wanted to give back. I remembered so vividly my dreams of being Chuck Yeager flying over Lancaster and breaking the sound barrier, and my pride in wearing the uniform of a Law Enforcement Explorer. One thing I learned during college was that Yeager was not the first choice to break the sound barrier—this honor belonged to a man by the name of Slick Goodlin. The problem with Goodlin was that he demanded a fee of $150,000—a huge sum of money in 1947—to fly the plane. Yeager, however, didn't want to do it for the money. He wanted to break the barrier out of a quest for adventure and a spirit of discovery. He wanted to see just what man was capable of achieving when pushed to his limits. Even with two broken ribs and so much pain he had to jerry-rig a broom handle to help him close the hatch of the plane, he would not be deterred.

Who was I? Was I the guy Oscar Wilde described, the one "who knew the cost of everything and the value of nothing"? I spent a good deal of my life trying to reconcile my inner Slick Goodlin and my inner Chuck Yeager. I had empathy for others who had struggled like me, who were in pain, and I wanted to help them. But I also wanted success. Practicing Ruth's magic had gotten me this far, and I continued to practice daily, knowing that I was only part of the way to where I wanted to be. I wanted fame and fortune. I wanted to be someone that others looked up to. I wanted to be the best surgeon in the world.

The army agreed to pay my way through medical school, all tuition and expenses, and I agreed to serve in the army as a doctor. I served a total of nine years in the U.S. Army, and eventually became Major James Doty.

MY MEDICAL SCHOOL experience was nothing like my undergrad experience. I had no difficulty academically, and I discovered I had a natural aptitude for studying the intricacies of the human body—anatomy, histology, physiology. The ability to memorize more information than it seems humanly possible for one person to memorize is the struggle for every first-year med student. I know now that my years of practicing what I learned in the magic shop had developed my brain so that I had the ability to memorize more easily than many of my fellow students. I could focus for much longer periods of time on my studies, and I never complained about my mind wandering off while I read medical textbooks. We were given mnemonics to help us remember everything from bones to nerves to how to write medical charts. Some were silly, like the mnemonic for remembering which cranial nerves are sensory, motor, or both—*Some Say Marry Money But My Brother Says Big Brains Matter More.* Other mnemonics were harder to remember than the original information, like OOOTTAF-VGVAH for the actual nerves of the cranium.

I used some standard mnemonics, other times I made up my own, and still other times I pretended I was using them when really it was just that the information I studied seemed to simply flow into my consciousness when I needed it. A 2013 study by researchers at the University of California, Santa Barbara, found that focused-attention meditation training improved memory, focus, and overall cognitive function in undergraduate students after just two weeks of practice, as measured by improved GRE scores and other memory and focus tests. What's amazing to me about this study is that the practices implemented by the researchers in 2013 were remarkably similar to my practice with Ruth in 1968. How much money is spent on GRE test preparation and courses? The beautiful thing about a meditation practice as a study aid is—it's absolutely free.

The army scholarship guaranteed me an internship after medical school but not a residency. In the civilian world, those two things are connected, but I would have to apply for a residency. After finishing at Tulane in 1981, I accepted a flexible internship at Tripler Army Medical Center in Hawaii—a place where I had done a prior rotation as a student. A flexible internship meant I would be focused on various surgical specialties rather than just general surgery. I did rotations in pediatrics, obstetrics, gynecology, internal medicine, and also general surgery, as well as neurosurgery. I thought this broad and varied experience would be more beneficial to my education, but what I didn't realize was that if you do a flexible internship it puts you at a disadvantage when you apply to a general surgery residency because you haven't focused on just surgery and its subspecialties. A broad knowledge of a lot of areas actually hurt my chances. My plan was still to become a plastic surgeon for children, which called for a general surgery residency, followed by a plastic surgery fellowship, followed by a craniofacial fellowship. I had a plan. But there were twelve of us competing for the general surgery residency, and I was the only one doing a flexible internship. The odds were not in my favor. My eleven colleagues told me there was no way in hell I would get into the general surgery residency, and they were clearly happy for my disadvantage. I had an intensity of purpose about me that did not go over well with the others, and my great belief in my ability to manifest anything I wanted was coming across as arrogance. I understand now why they seemed to want me to fail.

You apply for your residency in November, so I applied for the general surgery residency like everyone else. In April, however, I had my neurosurgery rotation. The guys who were in charge were the nicest I had met on any rotation. Neurosurgery was fascinating—working on the brain was demanding and precise, plus it gave me a thrill that I hadn't found in general surgery, which is mostly about the chest and abdomen.

There was something about going where no one had gone before, into the deepest recesses of what makes us human, that called to me. I still wanted to help children with deformities, but exploring the mysteries of the brain felt like a new quest that called to me. I wanted to be a neurosurgeon in the same way I wanted to go to college and to medical school, but to do that I had to do a neurosurgery residency, not a general surgery residency. I knew that I could do neurosurgery and still do a plastic surgery and craniofacial fellowship, if I wanted. It was perfect.

The chief of neurosurgery at Tripler was encouraging.

"You're technically very talented, Jim. You should do neurosurgery. You *need* to do neurosurgery."

"This is great," I replied. I was swollen with pride. I was going to be a neurosurgeon.

"The thing is," he added, "they train only one neurosurgeon a year in the army, and there's a backlog of three years. You're going to have to wait, and after your internship, they'll send you out in the field as a general medical officer for a few years until you are at the top of the list and can start your residency."

"Three years?" I asked.

"Only three years."

"I'm sorry, but I can't accept that."

He laughed at me. "You have to do your time, Jim."

"That's bullshit and unacceptable," I said more passionately and clearly out of line.

"That's just how it goes. It's not bullshit. It's the army."

"But it's unacceptable to me," I said.

He shook his head and showed me out of the office.

I had vacation time coming up, thirty days away from the army, so I left Tripler and went to spend a month at Walter Reed. This was where I planned to end up, so I rotated in neurosurgery on my own time and did very well. I met with the chairman of neurosurgery before my "vacation" was over.

"I like you, Jim, you've done an outstanding job during your rotation here, and I think you'd be an excellent resident."

"Thank you," I said. "I assume that means I'll start in the fall."

"Jim, you know that there is a minimum three-year wait. I will put you in at the end of three years. You should be thankful, as I already have four people who want that slot. You haven't even formally applied anyway."

I looked him directly in the eye and said, "Waiting three years is not acceptable. If you don't accept me next year it'll be the biggest mistake you'll ever make. I'm not waiting for three years. I'm sorry, I don't mean to be rude or brash, but I just can't accept that."

Even though it was late, I applied for the neurosurgery residency. I believed in the power of my own magic.

I went back to Tripler and told the chief of general surgery I was grateful for his consideration but I was withdrawing my application for general surgery because I would be doing neurosurgery at Walter Reed. "Impossible, you won't get in" was his official reply. "I'm not allowing you to withdraw. This is the best group of applicants I've ever had to this program, and you're one of them. I'm not letting you go."

"OK," I said, "but I'm just telling you I'm not going to do general surgery residency, and I will be at Walter Reed."

I finished out my internship, visualizing my neurosurgery residency at Walter Reed. Every morning and every night, I saw myself there in my mind's eye. I wasn't worried about the outcome, I had learned to visualize what I wanted and yet detach myself from the end result. It would happen, one way or another. That's all I knew. I did my footwork and trusted the details to unfold however they were meant to unfold.

As it turned out, the details were a bit salacious. The guy who had been accepted that next year had begun a relationship with a nurse at Walter Reed. The two broke up, and he began stalking her. Apparently there were other issues involved as well, and the chief of neurosurgery revoked the residency offer. He was reassigned to spend the rest of his time in the military as a general medical officer in South Korea. There was no backup for the position, as the other individuals who were slotted for neurosurgery residencies in the future were committed to complete their assignments elsewhere. As the dominoes fell, it turned out that I was suddenly the only person still standing.

I don't know if it was a result of my visualizing, a lucky series of circumstances, or something else. All I knew was that, once again, everything had worked out.

I received my letter of acceptance from the general surgery program at Tripler and from the neurosurgery program at Walter Reed on the same exact day. The head of general surgery had accepted four of us, and the day our letters came he brought us into his office.

"I want the four of you to know that each of you was my first choice for the four slots here at Tripler, and that this was the best intern class I have ever seen."

I looked at the other three who had also been accepted. They had gone out of their way to flatter the chief of general surgery who was also the chief of surgery. They had made sure that they had regulation haircuts and that their shoes were polished. Such was never my concern. I wanted to be the best intern that I could be and more often than not my hair was too long and my shoes were not shined. And I was never good at kissing ass. "I'm going to take you to the officer's club, and we're going to celebrate."

I interrupted the celebration and congratulatory pats on the back. "Sir," I said, "I want you to know I can't accept the position."

He looked at me. "Why the hell not?" he asked. No one ever refused once they were accepted.

"I've been accepted to neurosurgery at Walter Reed."

His face turned red. He was speechless. "I tried to warn

you," I said. "I told you to withdraw my application." I stood up, saluted him, and walked out.

THE CHAIRMAN at Walter Reed had told me he liked me during my monthlong rotation, but I turned out to be nothing but trouble for him. I was quick-witted and could use my tongue as a weapon. At Walter Reed I did this often. I felt compelled to stand up and tell the truth, no matter what, and this blunt honesty did not help my cause much as a resident.

I had become arrogant. The process of getting everything I wanted, and my technical expertise in neurosurgery, made me feel important and special in a way I never had before. The magic I learned at twelve, and had practiced now for over a decade, made me feel invincible. I got in trouble frequently. I had not yet learned discretion or discernment. I was confrontational with my chairman and often in front of others. Even as a junior resident I took being a doctor very seriously. I cared about my patients more than I cared for the pecking order and politics of residency. But my attitude alienated my superiors, and my chairman ended up disliking me intensely because I refused to follow any rules I didn't like or think were logical. I didn't care for the way the faculty and many of the senior residents bullied and belittled the residents, myself included, and it reminded me far too much of my childhood in Lancaster. I knew how to stand up for myself, and how to stand up for others, and I did so at every opportunity.

Right before Christmas, during my first year of residency, I was called in for an evaluation. The chairman was at his desk, and all the attendings were in the room.

"We'd like to go over your evaluation," the chairman began. "We have serious concerns, and there have been questions about how you take care of patients."

I immediately stood up and said, "Stop right there. If there are questions about my medical care, I want to see the documentation. I take being a doctor seriously, and I won't accept such accusations without proof." I had spent too many years watching my mother be ill-treated by doctors who didn't care. I had seen her dismissed. My family dismissed. I knew how much I cared for my patients. I listened to their stories. I double-checked everything having to do with their care. I came in after hours to sit at their bedside. I knew he was wrong.

There was nothing but silence in the room. The chairman began shuffling some papers on his desk, awkwardly.

"W-Well," he stammered. "It's not really about that. It's actually about your attitude. We don't think you really want to be here because you're confrontational, and we've decided to put you on probation. We're going to evaluate you for the next six months. If you don't perform, we're going to terminate you from the residency."

I looked from one face to the other. No one would meet my eyes.

"If you want to throw me out, throw me out. Right now. Probation is unacceptable. I will not do it. I've never been on probation for anything in my life, and I am not going to start now."

They were speechless. They couldn't terminate me, and I knew that they knew it. To do so would have been difficult, as all my evaluations from patients and the faculty were outstanding. It was only the chairman who had given me a negative review. Plus, it would have been a very big embarrassment.

"Wait outside and we'll call you back in after we've made a decision."

I sat outside the office for an hour and a half. I closed my eyes and focused on my breath. I tried to keep myself calm and trust what Ruth had taught me.

When they called me back in, the chairman cleared his throat and made his announcement. "We've decided we're not going to put you on probation, but we're going to be watching you. Closely."

It took everything I had not to laugh out loud. They were already watching me closely, and while my attitude with my superiors was not good, my way with patients and my talent as a physician were beyond reproach. I was smug, and still believed not only that I was invincible but that the magic Ruth had taught me would never let me down. Now I can see that I had learned the mechanics from Ruth but had missed the heart of her teaching.

"Well, OK," I said. "That sounds like a plan."

I antagonized my chairman for years. I was an excellent neurosurgery resident. He knew it and I knew it. I was never on formal probation, but when I graduated he shook my hand, leaned in close to my ear, and said, "I just want you to know that this whole time you have been on probation in my mind."

I had no humility, and my success in a white coat was going to my head.

Residency was serious business, but when we had breaks, it was an all-out party with no thoughts about the consequences. I worked hard, and I partied hard. I felt indestructible. Invincible. Just like I had imagined for so many years, I was wearing a white coat. I was Dr. Doty.

Nothing could stop me.

Residencies in the mid-1980s were even more grueling than they are now, a kind of medical boot camp—with as much as twenty-four hours at a time spent on shift. We were sleep deprived and under constant scrutiny and pressure. It became normal to blow off some steam now and then—take a break from the mental and physical demands of residency. Some of my colleagues began drinking more than they should—I recognized the signs in them and also in myself. I knew what alcoholism looked like from growing up, but I was trying to balance on the razor's edge between drinking too much on occasion and alcohol abuse. Partying on my rare time off wasn't being out of control, I told myself. I could feel the genetic pull in me at times to seek escape from the pressures and demands of life as a resident, but I wasn't my father. I would never be my father.

Gradually I stopped meditating and visualizing. Working long shifts didn't leave me with the time to practice every morning and evening. At first I started missing every few days, then I practiced only once a week. Until finally I felt there wasn't time at all. I had stopped adding things to my list. I knew exactly what I wanted, and I also knew just how close the grand finale to my magic show was. I was about to become a neurosurgeon, one of the elite specialists entrusted to operate on the most important part of the human body. The brain ruled all, or so I believed, and I ruled the brain. There was nothing more to learn from Ruth's magic.

One evening four of us decided to go out and celebrate the end of a particularly grueling rotation. We were a close group. We worked together, ate together, guzzled coffee in the cafeteria together. We had bonded the way people do when they go through a traumatic event or a natural disaster together. We were all fighting side by side in the same war—residency. Because we had no time for anyone else in our lives, by default we had become best friends, and a family, of sorts.

The pressure was extreme, and our way of relieving that pressure was also extreme. Working in the hospital, you see things that you wish you could unsee, and we found that the magic formula to blur these images in the mind involved a mix of large quantities of alcohol, cocaine, loud music, and half-naked women. Not necessarily in that order.

That night we started drinking around 8 P.M. at a strip club near the hospital. We threw money at the dancers as if we were guys who actually had money to throw away. We moved on to a Spanish restaurant where we ate paella and jamón serrano, a kind of cured pork served on toast. We drank jug after jug of some kind of Spanish wine. I'm not sure when the cocaine came out, but after pulling antique swords off the wall of the restaurant, and engaging in a life-and-death duel, we were all summarily kicked out.

It was a damp night in October, and as we left the restaurant, I remember turning my head into the mist and feeling its cool wetness on my cheeks. It felt good to be free of the hospital. It felt good to be alive. It felt good to be me. It felt good to be high.

The four of us piled into the car, which was littered with empty beer cans. We careened through the dark night with the music blaring. I felt myself drift into a happy stupor. Then I heard a voice in my head that said, "Put on your seat belt. Now!" I jolted alert and looked around. My buddy in the front seat was singing loudly and tossing beer cans out the window. My buddy who was driving was nodding his head in time with the out-of-tune singing. My buddy next to me, in the backseat, was asleep. None of them had told me to put on my seatbelt.

The car was a 1964 red Ford Fairlane—a classic that belonged to a friend's mother. None of us knew the tires were almost bald. There were lap belts in the backseat, and I reached for mine just as we hit a sharp curve in the road. The car started sliding and skidding across the wet asphalt, sideways, into the oncoming lane. I felt the seat belt tighten as centrifugal force pushed me against it, and then as if in a dream, I watched as we crashed head-on into a large tree.

Then everything went black.

The moaning brought me back into consciousness. I was lying on the wet pavement by the driver's side of the car. I don't know if I had been thrown out of the car or if my friends had dragged me out. My buddy who was driving was leaning over the steering wheel not moving. I felt a searing pain shooting up my back, but my legs felt numb. I tried to move them, but they weren't cooperating.

I began to vomit and tried to get up. I heard my buddies talking. *Rock Creek Park. It's a mile away. One of us has to go. My knee. You stay with him.* I couldn't piece together what it all meant, and I closed my eyes and let the wet pavement cool my face. My body was on fire, but I somehow believed that if I kept my face cool, I would be OK.

Walter Reed was only a mile away, so my friend from the backseat, who had only minor cuts and abrasions, set off on foot to get help. Once at Walter Reed, he told the staff they needed to dispatch an ambulance to pick us up. They refused, saying they had no authorization to attend to accidents off the base.

Undaunted, he requisitioned a government vehicle without authorization and drove back to the site. I screamed in pain as he dragged me into the backseat and delivered me to the emergency room. It was surreal to be examined by my fellow residents in the emergency department at Walter Reed. Hours before, we had been the doctors, but now we were the patients. My friends had torn ligaments, cuts, and one a pretty severe chest contusion and concussion, but in general they were OK.

I was the only one wearing a seat belt, and I was the one with severe injuries—a transection of the small bowel, a ruptured spleen, and a spinal fracture in the lower lumbar area. The abdominal injuries required immediate attention, and I was rushed into the operating room.

I had become a patient, and as I saw the operating lights shining down on me, it was as if I could feel what every surgery patient in that room had felt before. I felt the waves of pain, and fear, and worry. I heard voices. It was like being in a room full of people talking all at once. *What if I don't wake up? Please, God. Don't let it be malignant. I should have told him I loved him one more time. What if I never walk again? What will they do without me? Please. Help. I don't want to die.*

The next voices I heard were arguing. I opened my eyes

and could see I was in the intensive care unit. The pain was severe, beyond anything I had ever imagined. My stomach was bandaged. I closed my eyes against the light, and listened to the chairman of the General Surgery Department and the vice chairman of neurosurgery arguing. The argument was about me.

It wasn't good. Even through the pain, my medical education kicked in. My blood pressure had dropped precipitously since surgery. It was so low that there wasn't even a diastolic pressure being recorded. My systolic pressure, the higher number in a blood pressure reading and what measures the pressure in the arteries when the heart beats, was only forty. My blood pressure should have been at least two to three times that. My heart rate, however, was over 160. It was clear—I was in shock from blood loss. But I was still losing it and losing it rapidly, an indication of internal bleeding. Soon there wouldn't be enough pressure to supply my critical organs. I knew what this meant. I was going to go into cardiac arrest shortly. My brain was going to die. I was going to die.

I thought to myself that this wasn't how my life was supposed to turn out. I wasn't supposed to die like this.

In the next moment, I felt as if everything shifted and tilted. I was suddenly looking down at myself from the corner of the ceiling. I didn't feel any pain. I could see the rays of light coming off the lightbulbs in zigzag patterns. I could see every droplet of liquid in the IV bags. I could see the top of the chairman's head, and the sweat that dotted his forehead. I looked down and saw myself in the bed. I looked small and vulnerable, and very, very pale. I could see the monitors, their lines and numbers moving up and down erratically, and it seemed as if I could hear the blood in my vessels moving, and could sense that there wasn't enough. I could hear my heartbeat. It sounded like a far-off drum, pounding out a rapid rhythm. I observed all this without emotion. I didn't feel sad, just acutely aware of everything that was happening to me and around me.

The chairman of general surgery was insisting that he couldn't possibly have missed a bleeder in the abdomen, and this could not possibly be the source of my blood loss.

"You obviously missed something," the vice chairman was yelling. "He's oxygenating and has no major fractures. He's bleeding into his belly. You obviously missed a bleeder."

It was like watching a play, and at the same time I could feel both the frustration and fear of the vice chairman, and the pride and certainty of the chairman. I could feel what everyone in the room was feeling.

I saw the vice chairman put his hand on my leg. "You idiot, if you don't take him back into surgery, I'm going to. Now!"

Finally the chairman agreed. I watched from above as they wheeled me back into the operating room. One of the nurses leaned in and whispered in my ear, "Stay with us, Jim. We need you. You're going to be OK."

And then blackness.

My experience after this blackness is something that I could never adequately explain nor ever forget. It is all the more puzzling for being a rather common and yet extraordinary experience. One that has been repeatedly reported over the centuries.

Suddenly I was floating down a narrow river. I was moving slowly at first. Ahead I saw a bright white light, much like the tip of the flame I used to stare into at the magic shop. I began to speed up, and soon I was rushing toward it. All along the sides of the river I saw people I had known, crowding along the banks of the river. I thought I saw my father. I thought I saw Ruth. I felt loved and accepted in a way I never had before. Many of the people I saw were still alive. I saw my mom in her bathrobe. My brother laughing with me from our bedroom in Lancaster. I saw the girl Chris who I had a crush on in junior high school. I saw my old orange Sting-Ray bicycle. I saw myself on the bus to Irvine, and I saw myself trying on a white coat for the first time. I saw myself turning my face into the mist on that very same night. I felt the white light getting warmer and closer. It was getting bigger. I somehow knew that this light was love, and it was the only thing that meant anything in this universe. I just had to reach it, and I knew that when I did, I would be one with all things. This is what I had been searching for. This was the only thing I needed. I wanted to merge with the light. And suddenly I realized that when I merged with that warm, inviting light, I would no longer be part of this world. I would be dead. "No," I screamed. Or, at least, I thought I screamed. And suddenly I was going backward, away from the light. As if I had stretched a rubber band to its maximum and let go. I was going in reverse so fast that I could barely comprehend it. I felt the presence of all those who had greeted me now falling away.

My eyes were still closed, but I could hear the beep of the monitors.

I just had to open my eyes.

"Jim, can you hear me?" I felt a prick on my foot and opened my eyes. The bright light of the recovery room shined directly into my face, and I blinked rapidly.

"Jim!" the voice said. "I told you we needed you around here. Who's going to make us laugh and catch all the heat if you're not here?"

I reached my hand out and touched her arm. "Am I alive?"

"Of course you're alive. We had to pump a lot of blood into you, but you're going to be fine. You're stable."

"Are my friends OK?"

"They're fine. You guys are all lousy patients, but you'll be fine. Unless we kill you in your sleep." She laughed.

"Did I die?" I asked.

"You're alive."

"No, I mean, did I die? Did I have to be resuscitated in there?"

"No. You were pretty unstable and your blood pressure was

really, really low, but you didn't have a cardiac arrest. They found a bleeder they had missed near your spleen. You had four liters of blood in your belly. No wonder your pressure was low. They had to transfuse you sixteen units. But no, you weren't dead. At least, not that I know of."

She looked at me questioningly.

"It's nothing. It was just weird. I was on a river." I stopped talking then. Whatever that experience was, I had no need to explain it. The scientist in me started reviewing the physiology of the event. Could my experience be an extreme form of low oxygen to my brain? Had I had a massive release of neurotransmitters? Was it all a hallucination due to shock and trauma and blood loss? While I was in this experience, I wasn't a neurosurgeon looking on it with medical knowledge, but now I was. Was this a mystery of the brain I could ever solve?

IT'S ESTIMATED that up to fifteen million Americans have had a near death experience, or NDE, as they're commonly called. In 2001, the journal *Lancet* published a study showing that between 12 and 18 percent of patients who experienced cardiac arrest or cessation of breathing might have had near death experiences after medical conditions involving low blood pressure, impaired brain oxygenation, or global impairment of brain function through trauma or disease. These similar experiences often include descriptions of being out of one's body, floating, a flashback of one's life, having a feeling of being with deceased loved ones or hearing their voices, a feeling of warmth and unconditional love, and often traveling on a river or being in a tunnel while being drawn toward a light. Such descriptions have also been described in multiple cultures and throughout recorded history.

In Plato's *Republic* there is the "Story of Er," in which a soldier has been slain, is found not to decay, and awakens on his funeral pyre twelve days later. He gives a similar description of his own near death (or death) experience, including several of the common elements associated with modern NDEs. Some have claimed that the famous sixteenth-century painting *Ascent into Empyrean*, by the Dutch artist Hieronymus Bosch, is a representation of a near death experience with its tunnel that leads to a bright light with shapes and forms possibly representing the world beyond life on earth. There is also the story of British admiral Beaufort, who described his near drowning in 1795, and American physician A. S. Wiltse, who in 1889 described his own similar experience during a bout of typhoid fever. Each of these descriptions has several components associated with classic NDE—seeing their body from a distance, a sensation of floating, seeing loved ones, and traveling toward a white light.

In the late nineteenth century, Victor Egger, a French epistemologist and psychologist, used the French term *expérience de mort imminente* (experience of imminent death) to describe a similar phenomenon occurring in climbers who "saw" their lives pass before them as they fell to what they thought would be their deaths. More recently, Celia Green, in 1968, published an analysis of four hundred accounts of out-of-body experiences that led people to question whether our consciousness can exist out of our bodies, and in 1975, psychiatrist Raymond Moody published a book of such experiences and coined the term *near death experience*, garnering the interest of scientists in this phenomenon, which previously had been described only in the realm of religion, philosophy, and metaphysics. Many descriptions include religious symbols like angels and figures such as Jesus or Muhammad. Usually such symbols correlate with the faith or religious beliefs of the individual. For many, such experiences are life altering. Individuals who are atheists report many of the common NDE elements as experienced by believers. One of the most famous is that of Sir A. J. Ayer, a British philosopher and the author of *Language, Truth, and Logic*, an avowed atheist, who in 1988 almost choked to death while eating. Following the event, he was quoted as saying, "My experiences have weakened, not my belief that there is no life after death, but my inflexible attitude toward that belief." Among the recorded NDEs of atheists, a number report no impact on their beliefs, while for others there has been a spiritual conversion.

Because of the work of Moody and others there is a growing interest among scientists to study this phenomenon. In addition, we know that similar experiences can be artificially induced through such medications as the anesthesia drug ketamine and some psychedelics. They can be triggered by electrical stimulation of the temporal lobe or hippocampus in the brain. They can happen during decreased levels of brain oxygenation through decreased blood flow to the brain (as experienced by fighter pilots) and even during hyperventilation. It's interesting that, while induced experiences have components of the NDE, with the exception of psychedelics, they are not typically associated with transformational or life-changing responses in the individuals who experience them. Is it truly the risk of death (or a part of the brain that interprets the situation as such) that is the common denominator in these situations that makes them transformational?

It has been postulated by psychologist Susan Blackmore that the experience of passage down a tunnel toward a bright light is a result of increasing neural noise occurring as more and more brain cells start firing in response to a lack of oxygen to the brain. She also suggests that the sense of serenity and peace is due to a massive endorphin release from the stress of the event. In a recent study, physiologist Jimo Borjigin, using a rodent model of hypoxia, demonstrated a transient surge in synchronous coherent gamma oscillations, which were global and highly coherent, occurring within thirty seconds of cardiac arrest. In other words, rats deprived of oxygen and who go into cardiac arrest and die had brains that showed a heightened consciousness after death. These gamma oscillations are noted in both wakeful consciousness and heightened states of consciousness associated with meditation as well as during rapid eye movement (REM) sleep, which is the period during sleep when memories are consolidated and strengthened. Clearly, there are a number of well-documented neurophysiologic events that are occurring during NDEs and that can occur during other types of brain stressors or be replicated utilizing a variety of methods not associated with an NDE.

Like so much of life, our beliefs are a manifestation of our life experiences. And our brains are the consolidation of those experiences. But what about the experiences of the heart? Even more interesting to me than the science, the research, and the questions about an afterlife that result from a near death experience is the common thread that runs through these experiences. Why is it that so many travel toward the light and the warmth and the love? Perhaps what we experience during NDEs are our heart's greatest longings. To be loved unconditionally. To be welcomed. To feel the warmth of home and family. To belong.

I don't know exactly what happened to me after that car accident when my blood pressure dropped precipitously low, and in the end, I realized that it didn't matter. I didn't need to solve it or explain it. Maybe I died, maybe I didn't.

I just don't know.

What I do know for sure is that I have died many times in

this life. As a lost and hopeless boy, I died in a magic shop. The young man who was both ashamed and terrified of his father, the one who had struck him and got his blood on his hands, died the day he went off to college. And although I didn't know it at the time of my accident, eventually the arrogant, egotistical neurosurgeon I would become would also suffer his own death. We can die a thousand times in this lifetime, and that is one of the greatest gifts of being alive. That night what died in me was the belief that Ruth's magic had made me invincible and the belief that I was alone in the world.

At the time I felt the warmth of a light and a sense of oneness with the universe. I was enveloped in love, and while it didn't transform my religious beliefs, it informed my absolute belief that who we are today doesn't have to be who we are tomorrow and that we are connected to everything and everyone. I woke up in that hospital bed, and I remembered just how far I had come from that orange Sting-Ray bike and a summer spent in a magic shop. What I didn't know then was how far I still had left to go. Seeing Ruth along that river, feeling love and connection to so many, was perhaps a warning sign that I was straying too far away from what she had been trying to teach me. But it would be many more years and many more painful mistakes before I realized it.

NINE

The Sultan of Nothing

Newport Beach, California, 2000

One morning I woke up worth $75 million. I didn't actually have this money in hand. In fact, I'd never seen it or counted it, but it existed in a place even more powerful than any bank—my mind.

I was single, having already been married and divorced by this time. The long hours of being a neurosurgeon and the pursuit of wealth and success hadn't made me a very good husband or a very good father to my daughter. The divorce rates among physicians are said to be 20 percent higher than those of the general population, and the rate for the neurosurgery profession is even higher. I was no exception to this rule.

I reached my arm across the bed until my hand landed on the warm body next to me. Her name was Allison, or maybe it was Megan. I couldn't remember exactly, but her skin felt warm and smooth and soft. I heard her murmur as she rolled onto her side.

I quietly got out of bed and headed downstairs. I needed coffee, and I needed to check what the stock market had been up to while I was sleeping. I turned on the computer and waited while it hummed and cranked to life. I was forty-four years old, and my plan was to retire within the next year. My life in Newport Beach was a long way from Lancaster. I had become one of the most successful neurosurgeons in Orange County. I lived on a bluff overlooking Newport Bay in a seventy-five-hundred-square-foot home. My garage held not only the Porsche I had dreamed of as a boy but a Range Rover, a Ferrari, a BMW, and a Mercedes.

I had everything on my list and more—much more.

A few years earlier, a friend of mine had shared with me his idea for a technology that would revolutionize the field of radiation therapy and the treatment of solid tumors in the brain. He had just finished his residency and accepted a position at Stanford, where he planned to develop the concept from simply an idea to a reality. He had started a company. I was so impressed that I became one of the first investors. I told him that I would place the first unit outside Stanford down in Newport Beach. Little did I know that one interaction would change the trajectory of my life. I did place the first unit of the newly named CyberKnife in Newport Beach. I convinced another physician friend who had significant family wealth that this technology would change the world. He believed me, and not only did he buy the first unit but he also bought a building to house it and MRI and CT scanners to use with it. He spent millions of dollars based on my enthusiasm and belief in the technology. At the time, the device had yet to be approved by the FDA and there were no codes that one could use to bill for its use. Within two years after his investment, Accuray, the manufacturer, was effectively bankrupt through a combination of mismanagement and an inability to raise adequate capital. Several years later, they still had not been able to get FDA approval, and sales were nonexistent. The company had burned every bridge, not only in Silicon Valley but in the entire United States in terms of raising further capital. Things looked grim, and those who had faith in the potential of the technology and who had put millions of dollars into it were going to lose their investment and the world would lose this extraordinary technology. I had to do something. I decided I was going to save the company.

I HAD no significant background in business, although as a resident, I had invented an electrode for monitoring brain activity that was being sold throughout the world. This was different, though. This was the big time. I told my friend that I had a plan. I'm not sure if he believed that I could help or if he had no other options, but nonetheless, he encouraged me. The company had gone from sixty employees to six. I agreed to fund the company myself while figuring out how to save it. I had no idea what I was going to do. As fate would have it, the answer came while having a drink at the bar in the Four Seasons Hotel, which at that time was located in Fashion Island in Newport Beach. I was waiting at the bar for a woman I was to have dinner with and began a conversation with the fellow sitting next to me. I shared with him the situation regarding the CyberKnife and how the technology could save hundreds of thousands of lives. I just needed someone to raise the money necessary for it to survive. He ended up helping me restructure the company and raise $18 million. The problem was, the principal investor would agree to do it only if I would agree to be the CEO. I not only had sold them on the concept but on me as the critical component to its success. So I walked away from my very successful private practice in Newport Beach to be a CEO. A job for which I had no experience and no expertise. The only thing I had was an absolute belief that I could save the company and that I had to save the company.

Within eighteen months the company was completely re-structured, FDA approval was obtained, and the valuation had gone from effective bankruptcy to $100 million. During this time, I ended up meeting a lot of people, including venture capitalists and entrepreneurs who were starting companies in Silicon Valley. They all thought I had some sort of secret magic to turning around Accuray and making a failure into a success. I didn't. I tried to tell them I knew nothing, but more often than not, they asked me to invest or become a partner in their firm or at least consult with them. And those investments and relationships resulted in me getting stock. Lots of stock. And in 2000, when the dot-com boom was at its zenith, publicly held stock in a dot-com was better than gold, and it guaranteed a line of credit at any bank.

The computer finally came online and I checked my numbers. I was still worth somewhere north of $75 million. As a boy I had dreamed of making a million dollars, but the thrill of my first million was nothing compared to my thrill at being worth $75 million. I was rich. I shut the computer down and looked out the window at the expanse of blue Pacific.

The house was quiet. Megan or Allison hadn't woken up yet, but I didn't want to share my news with her. Just the thought of her made me feel a little sad. We didn't know anything about each other. I knew she was a pharmaceutical rep, and she knew I was rich and had my own private table reserved for me at the best restaurant in Orange County. She had approached me last night with a group of her friends. We had drunk vodka and champagne, and when I had asked her what she thought of all this crazy excess, she had just laughed and said she thought I was great. I knew she had a story, but she wasn't willing to share it with me, and she didn't seem all that interested in hearing mine. And so, like many other nights with many other women, we had both agreed to feign an intimacy that didn't exist. We shared our bodies, but didn't let our minds or hearts complicate things. It made me feel lonely and hollow, but I had learned how to ignore the voices of doubt and despair in my head a long time ago.

I had everything I had ever dreamed of having. People respected me. People deferred to me. I had just agreed to purchase a private island in New Zealand and had wire-transferred the down payment. I owned a penthouse in San Francisco and a villa in Florence overlooking the Ponte Vecchio. I had wealth beyond my wildest imagination, accomplishments that I would put up against anyone else's in medicine or business, but loneliness was an indulgence I couldn't afford.

My plan was to retire and spend part of my time donating my medical services in third world countries and the rest of my time traveling between San Francisco, Florence, and New Zealand. If it felt like something was missing, I didn't worry about it too much. Whatever it was, I would find it in my travels.

Allison or Megan found her way downstairs and we stood around awkwardly and waited for the cab I had called to come pick her up. I had a meeting with my lawyers and then I was heading to New York for the week on business. I promised to call her when I returned. She wrote her number down on a piece of paper. After a dry kiss good-bye, she left, and I picked up the piece of paper and put it in a kitchen drawer. She had written her name above the number. It wasn't Allison or Megan. Her name was Emily. It didn't really matter. We both knew I had lied when I said I would call.

THE TWO ATTORNEYS graciously ushered me into their offices. An investor friend of mine had recommended this law firm to me because apparently they were rumored to handle the sultan of Brunei's U.S. holdings. I didn't know if this was true because their clientele was supposed to be kept confidential. My accountant had advised me to set up an irrevocable charitable trust, earmarking certain holdings for charity in order to reduce my tax liability. This law firm would be drawing up the paperwork.

"We've reviewed your portfolio, Dr. Doty, and you have significant holdings," said the attorney. "There are various types of charitable trusts. Have you discussed these with your accountant? This is no small consideration for a man of your worth."

I absorbed his words. A man of my worth. I took a deep breath and heard the voice in the back of my head wondering just who I was really trying to prove my *worth* to—myself or the world?

"I have. He's advised me to create an irrevocable trust."

"And do you understand the legal ramifications of such a trust?" asked the second attorney.

"It's irrevocable?" I quipped.

Corporate attorneys rarely have a sense of humor. "In order to see any immediate tax savings, it must be irrevocable. This means that once you fund it, you can't make any changes to the trust or take any of the property back. In this case, we're talking about stock in Accuray."

I had decided to donate my stock in Accuray—it wasn't my most valuable stock, but it was potentially worth millions. I was planning on allocating the bulk of it to Tulane and some to Stanford, where I had become a faculty member and where the CyberKnife had been developed. By this time my brother had died of AIDS, so my plan was to donate some of the stock to fund HIV/AIDS programs, as well as various charities and programs that helped underprivileged children and struggling families. Part was to go to support health clinics in various parts of the world.

"I understand," I said.

"If you're uncomfortable with the permanency, you can always make it revocable until your death. That's an option some people choose, but the tax consequences are different."

"I'd like to make it irrevocable," I said. Giving this money was important to me. I wasn't going to change my mind.

"Very well," said the first attorney. "We'll draw up the papers." We spent the next two hours reviewing my stock and the charities I wanted to gift. By the end I felt important. Generous. And the lonely, hollow feeling I had woken up with was gone.

The sultan of Brunei had nothing on me.

I flew first class to New York City and checked into a suite at the Palace Hotel, which at the time was coincidentally owned by the sultan of Brunei. A good friend of mine managed the hotel, and his friendship resulted in their giving me a huge suite. The culmination of my week in New York was a meeting with a hedge fund manager who wanted me and another investor friend of mine to help him with a company he had funded in Silicon Valley. He was absolutely convinced that our involvement in his company would ensure its success. I had tried to dissuade him—saying I really didn't think we could help, but he thought I was just being overly modest. When I said that, my investor friend had kicked me under the table.

We were meeting about our potential partnership and also about the opportunity I had to put a collar on some of the stock I owned. The stock was worth tens of millions, but there were a few rumblings in the market that the boom couldn't last. By putting a collar on the stock, I would still be paid at a predetermined number that would safeguard against a market crash, and if it went up, it could still be bought at the predetermined price so the buyer would gain the upside. Several people had advised me to hedge my investments in this way.

We met at Le Cirque, an upscale restaurant then located at

the Palace Hotel. We drank bellinis and bohemian sidecars. The meeting was a formality, as we had already agreed that he would give us 50 percent of the company, and we would help raise further equity investment and give strategic advice. We discussed it briefly and then moved on to my desire to put a collar on my most valuable stock—Neoforma. After discussing and agreeing on the terms, he gave me some paperwork to look over and complete.

My friend, who had been sitting there silently but drinking heavily, suddenly blurted out, "We want sixty percent of the company."

Apparently the bellinis had given him some newfound knowledge of our ability or importance, and he decided that we needed to own the majority of the company.

"What are you talking about?" asked the hedge fund manager. "We agreed on fifty percent twenty minutes ago."

"If you want our expertise, it's sixty percent, or forget it." The alcohol had made my friend greedy and illogical. He was trying to take advantage of the situation, and I had no idea why he was doing this. I would have been happy with the deal at 30 percent, and I had told him so earlier in the day.

"We have a deal at fifty percent."

"If you keep talking, I'm going to make it seventy-five percent. Or maybe we will cut you out altogether." He was yelling now, and I could see the other patrons start to look over at us nervously.

"You're an asshole," the hedge fund manager said.

At that moment, everything exploded. The two of them jumped out of their seats, and I lunged between them before they could come to blows. People didn't usually get into screaming fights at Le Cirque, and I was mortified.

We left and I flew home the next day extremely pissed off at my investor friend and worried that I couldn't reach the hedge fund manager on the phone to apologize. I kept trying to reach him, only to be told that he wasn't in and I should leave another message with his secretary. There was no doubt he was trying to avoid me.

I paced around my home in Newport Beach. I had a bad feeling about the whole deal, and it took six weeks before the guy finally returned my call.

By then it was too late.

The stock market was crashing, and people were frantic. Stocks were dropping in value, people were losing millions, and although we wouldn't realize it or have a name for it until later—the dot-com bubble had burst.

My net worth had plummeted, and I read financial statement after financial statement confirming what I already knew to be true. The $75 million was gone.

Not only was it gone, but because of the lines of credit based on stock valuation, I was also several million dollars in debt, and effectively bankrupt.

The only tangible asset I had left, and the only stock that

was still worth the paper it was printed on, was the company I had saved from bankruptcy and rebuilt from the ground up—Accuray.

But that was in an irrevocable trust.

I was worth absolutely nothing.

Less than nothing.

IT SEEMED that all of my friends disappeared almost as quickly as the zeros in my bank account. There were no more free drinks, free meals, VIP seating in the best restaurants. It took almost two years of struggle—and after selling the penthouse, the cars, the villa, and canceling the purchase of the island, I still owed money. Month after month I watched everything I had worked so hard for go away. All the money, power, and success I had dreamed of and visualized in my head since I was a teenager was gone—vanished in one big pop of a bubble. I had made it appear and then it had disappeared.

"Don't worry," one of the few friends I had left said. "You can work that Doty magic again."

Was it really magic? All the start-up investing I had done, and the success that came with it, seemed like a fluke. I had gotten drunk on amassing a fortune and the power that came with it. But ultimately I was a neurosurgeon, not a technology guy. I had some skill at investing, and I was really good at making things happen and making people believe. I knew how to work hard and focus and how to think big and get others on board, which had made me wildly successful. At the heart of it all, however, my greatest strength was as a healer, not an entrepreneur.

I grieved the loss of my fortune and my lifestyle, and on the day I packed up my house in Newport Beach, I felt empty, lost, and more alone than ever. I had lost my marriage. I wasn't involved in my daughter's life. I couldn't think of a single person I could call and share how I was feeling. In the pursuit of things, I had neglected relationships. And when I needed someone the most there was no one there.

While packing up the house, I found my old box of special things in the back of a storage closet. I hadn't opened it since college. I took out my old notebook, and I opened up the page and read over my list of things I wanted out of life when I was twelve. There were other pages of writing—places where I had written down what Ruth taught me, and funny phrases she had said that at the time I hadn't really understood. Everything on my list had materialized, but now it was all gone.

I was a horrible magician.

I HAD DIVIDED my six weeks with her into four parts. *Relaxing the Body. Taming the Mind. Opening the Heart. Clarifying Intent.* In the margin above the third section I had written *moral compass* with a question mark after it, and *what you think you want isn't always what is best for you.* This had three question marks after it.

I sat on the floor in front of the closet in my almost empty house, and for the first time in a very long time, I took three deep breaths and began relaxing every part of my body. I focused on my breathing, in and out, inhale and exhale. I felt my mind quiet. Then I focused on opening my heart. I sent love to the boy I once was and to the man I had become. I opened my heart to the truth that I wasn't the only one who had experienced loss, and I opened my heart to all those who were struggling just to eat, to have shelter, to take care of their children. And then I visualized the window in my mind, and it was opaque. I couldn't see what was on the other side of the window—what was in my future—no matter how hard I tried. For the first time since I had met Ruth, I didn't have a vision for what I wanted next or who I wanted to be. I had no idea of what I wanted to be on the other side of the window.

In that moment, I knew what I needed to do. I had to go back to the magic shop—back to Lancaster. Maybe Neil was still there. Maybe Ruth was still alive. I tucked my notebook under my arm and grabbed the keys to my one remaining car. I had kept the Porsche. It was the first car I had dreamed of and I owned it outright.

Lancaster was only a few hours away.

I could be there before dark.

PART THREE

The Secrets of
the Heart

TEN

Giving Up

If my life had been a movie, I would have arrived in Lancaster to find Ruth waiting for me in the magic shop. Ruth would have been approaching ninety but would appear more wise than frail. She would have sensed I was coming and had some meaningful words that would help me make sense of my failures.

Life isn't a movie, however, and when I arrived in Lancaster and drove to where the magic shop had been, it was gone. The entire strip mall was gone. I called information and asked for a list of magic shops in Lancaster. There were no listings for magic shops. There was a listing for a magician in nearby Palmdale who did children's birthday parties, so I dialed the number.

"Hi, I'm looking for a magic shop that used to be in Lancaster," I said. "It was owned by a guy named Neil. I don't know the last name."

There was a pause on the other end.

"Are you looking for a magician?" the man asked.

"Yes, by the name of Neil. He owned Cactus Rabbit Magic."

"There's no one by the name of Neil here. I think you have the wrong number."

I tried to curb my frustration. "Did you ever go to a magic shop in Lancaster, by any chance?"

"There's no magic shop in Lancaster," he said with a slight annoyance to his voice. "You have to go to Los Angeles to find a good magic store."

"There used to be one. In the late sixties. I was just wondering if you knew anything about it or what happened to the owner."

"Well, I was born in 1973."

I sighed. This was not working. "Thanks anyway. Sorry to bother you."

"You know, I do remember hearing something about a magic store in Lancaster that closed down in the eighties. I think the guy made cards or something. Got pretty famous, but I can't remember his name. You might want to try the Magic Castle in Los Angeles. A lot of old guys hang out there."

I thanked him again and hung up the phone.

I set out on foot and realized I was tracing the same route I used to ride on my bike to and from the magic shop every day. Everything was different. Lancaster was more like a real city now, not the isolated desert town of my childhood. I walked past the still-empty field where I had run into the bullies and now saw kids playing and laughing. The church next door was also still there just as it had been. Some things hadn't changed. I walked all the way back to the apartment building we had lived in during that summer. It looked pretty much the same, just older and even more decrepit than I remembered. Our apartment had been on the ground floor and a bicycle was lying on its side on the porch just the way mine once had over thirty years earlier. I walked around the corner to the room my brother and I had shared. Torn curtains partially covered the windows, but I could see some figures on the window ledge, and I walked a bit closer over the yard that was more dirt than lawn. And there were Captain America and the Avengers. I remembered how I had used the same ledge for my own action figures, only mine were G.I. Joe, Captain Action, and the Man from U.N.C.L.E. I turned back to see the tree I used to climb sometimes to get away from my parents' fighting, sometimes to just be alone, sometimes to cry because I felt so alone. I walked a little farther into a field of tumbleweeds and junk and looked around. For a few seconds I just stood looking over the field. I felt like that kid again, and felt the excitement of jumping on my bike, heading to see Ruth. I followed the path through the field that I used to take. I was suddenly brought back to reality by a horn honking.

I realized I wasn't sure what I was looking for or even why I was in Lancaster. Ruth didn't live here. She was from Ohio, if she was still alive at all. I didn't even know her last name. I walked back to my car feeling like I was missing something important. What had I come here for? What was I *really* looking for?

My notebook sat on the passenger seat. I picked it up and started reading through my Ruth notes. *Compass of the heart.* It was underlined. I didn't remember it being underlined earlier in the day, but I must not have noticed. There were also stars I had drawn in red ink on either side of the words. I flipped through the rest of my Ruth notes. Nothing else was underlined nor were there any other stars. Why this phrase? I closed my eyes and tried to remember when Ruth had said this. It was the day of the fight. The only day I had ever been late. The day she had told me about opening my heart. I remembered sitting in the chair in the back room, the smell of the place, and then came the bits and pieces, like song lyrics or poetry.

Each of us in our lives experiences situations that cause pain.
I call them wounds of the heart.
If you ignore them, they won't heal.

*But sometimes when our hearts are wounded that's when
  they are open.*
*Frequently it is the wounds of the heart that give us the
  greatest opportunity to grow.*
Difficult situations.
Magic gift.

I opened my eyes. I remembered when I was leaving that day—Ruth had followed me out to the parking lot.

"Do you know what a compass is?" she asked.

"Sure," I said. "It tells you what direction to go in."

"Your heart is a compass, and it is your greatest gift, Jim. If you're ever lost, you just open it up, and it will always steer you in the right direction."

I read the other sentence in the top margin. *What you think you want isn't always what's best for you.* Ruth had warned me. She had told me to open my heart before visualizing what I wanted and to use the power wisely. I hadn't done that. Could I have gotten it all wrong? I thought I wanted money. But the truth was, I had gotten money, but there was never enough money to make me feel like I had enough. It was as if the magic show I had begun so many years ago had now stopped. I had just kept pulling out one trick after another, so the applause never stopped, and the show kept going, and the millions piled up. And I was still just as alone, and scared, and lost as I was the first day I met Ruth. If I were being completely honest, there was a part of me that felt completely free now that the money had disappeared.

No magic trick lasts forever.

I WOKE UP the next morning to the sound of the phone ringing. It was after 10 A.M. There was no woman in my bed, and I didn't have to get up early to check the stock market. I had fallen asleep visualizing my heart opening, and I had asked the compass of my heart to steer me in the right direction. Then I had slept soundly, better than I had in years.

One of my attorneys was on the phone, and he said he had some big news for me.

"What is it?" I asked.

"I was reviewing your trust documents and realized that it had never been formalized or filed and therefore never completed. For some reason, this was never done and I can't see any specific reason in the file why this was the case. It was just an error that was missed. The notes all document your intention and list how much stock for each charity. I checked with one of our senior partners, and he said that, based on these facts, you have no requirement to fund the trust or complete the documents."

I sat down on the edge of my bed. Had the magic worked just like it had the first time when the rent money came through at the last second? I sat on the edge of my bed, holding the phone.

"Jim, are you there? Did you hear me?"

"I heard you," I answered. "Thanks for calling."

"Well, how would you like me to proceed?" he asked, no doubt surprised that I wasn't jumping up and down like a man who had just won the lottery. I did not know how much the stock in the trust would be worth, but I knew I would have been a millionaire again. All I needed to do was nothing.

"I'll call you back," I said, and hung up the phone.

One of humanity's most enduring myths is that wealth will bring happiness and money is the solution to any problem. I had lost my money, and that was a problem. Now I potentially had a good chunk of it back, and that was also a problem. I had given my word to these charities. My father had been full of empty promises, and I had vowed to myself that I would never be a man who didn't keep his word.

I knew people would understand. No one would expect me to willingly give away every bit of my remaining wealth in my present situation. No one would fault me. In fact, the head of the giving offices at two of the largest charities told me people renege on significant donations all the time, even after signing documents. That's an accepted reality. People's situations change. My situation had changed. I was no longer in a position where I could just give away millions of dollars.

Or was I?

I closed my eyes and imagined my heart opening. I sent love and forgiveness to myself for all the mistakes I had made. I sent love out to my parents and gratitude for them doing the best they could. I sent love to Ruth, wherever she was, because she was the kindest person I had ever known. And I sent love out to every child who was struggling with poverty, or who had parents who were addicted, or who were alone and somehow thought it was their fault. I sent love to every person who had ever questioned their own value or worth, and to every single person who thought money defined them. I closed my eyes and opened my heart. I felt something I had only ever felt once before in my life—a feeling of being enveloped by warmth and love . . . a sense of deep inner peace and an absolute certainty that everything was going to be OK—only this time I wasn't going down a river toward a white light while bleeding out on an operating table.

I opened my eyes and picked up the phone to call the attorney back. "I'm going to sign the trust paperwork and donate everything as planned."

He said, "You're kidding, right?"

"No, I'm not kidding. Do it."

As I was hanging up the phone I heard him say, "Holy shit." And then there was just silence. I didn't have millions of dollars, but I was still a neurosurgeon. I wasn't going to starve. I was still going to be wealthy by any normal standards, but I wasn't going to have a fortune. It was time to start over and truly become a person of worth and value that had nothing to do with any dollar amount. This was what Ruth had wanted to teach a young boy, but some lessons can't be taught and have to be learned by experience to be learned at all.

I didn't know that, in 2007, when Accuray went public, it would be valued at $1.3 billion and my charitable trust would be worth $30 million. Even if I had known, I wouldn't have changed my decision. In that moment I felt free, free to follow the compass of my heart, and that was priceless. The monkey that had been holding so tightly to my back and had driven me with the false belief that money would make me happy, that money would give me control, suddenly let me go. I learned that there's only one way for wealth to bring happiness—and that's by giving it away. I was free.

The brain has its mysteries, but the heart holds secrets that I was determined to uncover. My quest that began in the magic shop had taken me on a journey inward, but my journey wasn't over. I knew I had to travel outward. The mind wants to divide and keep us separate. It will teach us to compare ourselves, to differentiate ourselves, to get what's ours because there is only so much to go around. The heart, however, wants to connect us and wants to share. It wants to show us that there are no differences and that ultimately we are all the same. The heart has an intelligence of its own, and if we learn from it we will know that we keep what we have only by giving it away. If we want to be happy, we make others happy. If we want love, we have to give love. If we want joy, we need to make others joyful. If we want forgiveness, we have to forgive. If we want peace, we have to create it in the world around us.

If we want our own wounds to be healed, we have to heal others.

It was time for me to focus again on being a doctor.

WHAT RUTH CALLED the compass of the heart is really a form of communication that exists between the brain and the heart through the vagus nerve. What research has shown is that the heart sends far more signals to the brain than the brain sends to the heart—and while both the cognitive and emotional systems in the body are intelligent, there are far more neural connections that go from the heart to the brain than the other way around. Both our thoughts and our feelings can be powerful, but a strong emotion can silence a thought, while we can rarely think ourselves out of a strong emotion. In fact, it is the strongest emotions that will trigger ruminating or incessant thought. We separate the mind as rational from the heart as relational, but ultimately the mind and heart are part of one unified intelligence. The neural net around the heart is an essential part of our thinking and our reasoning. Our individual happiness and our collective well-being depend on the integration and collaboration of both our minds and hearts. The training Ruth gave me would help integrate both brains in my body, the mind-brain and the heart-brain—but for decades I ignored the intelligence of my heart. I thought I could use my brain to lift me out of poverty, to lead me to success and give me value, but ultimately it was my heart that gave me true wealth.

The brain knows a lot, but the simple truth is it knows a lot more when it joins with the heart.

Mindfulness and visualization, the current names for what Ruth taught me, are wonderful techniques for getting quiet, eliminating distraction, and journeying inward. They can increase focus and help us make decisions more quickly, but without wisdom and insight (opening the heart) the techniques can result in self-absorption, narcissism, and isolation. Our journey isn't meant to be an inward journey alone, but an outward journey of connection as well. When we go inward, and our heart is open, we will connect with the heart, and the heart will compel us to go outward and connect with others. Our journey is one of transcendence, not endless self-reflection. There's a reason stock traders are using meditation techniques; these techniques help them become not only more focused but, sadly in some cases, more callous. This is what Ruth warned me about before she taught me to visualize. Yes, we can create anything we want, but it is only the intelligence of the heart that can tell us what's worth creating.

There is an epidemic of loneliness, anxiety, and depression in the world, particularly in the West. There is an impoverishment of spirit and of connection with one another. Studies show that 25 percent of Americans have no one that they feel close enough with to share a problem. This means that one out of every four people you see or meet today has no one to talk to, and this lack of connection is affecting their health. We are wired for social connection—we evolved to be cooperative and connected with one another—and when this is cut off, we get sick. Research has shown that the more connected we are socially, the longer we will live and the faster we will recover when we get ill. In truth, isolation and loneliness puts us at a greater risk for early disease and death than smoking. Authentic social connection has a profound effect on your mental health—it even exceeds the value of exercise and ideal body weight on your physical health. It makes you feel good. Social connection triggers the same reward centers in your brain that are triggered when people do drugs, or drink alcohol, or eat chocolate. In other words, we get sick alone, and we get well together.

By giving up my last remaining wealth, I learned the lesson I was too young to comprehend during my time with Ruth. The grand finale of the magic that Ruth taught me was the ultimate insight that the only way to truly change and transform your life for the better is by transforming and changing the lives of others.

Ruth taught me techniques and practices, but by taking the time to teach me, by giving me her time and attention, she taught me the greatest and most real magic there is—the power of compassion to not only heal each of our own wounds of the heart but the hearts of those around us.

It's the greatest gift, and the greatest magic.

ELEVEN

The Alphabet of
the Heart

Mississippi, 2003

Everything is beautiful at a distance. After returning to medicine, I could look back at my life in Newport Beach and see the beauty in every mistake, every wrong turn, and every misguided belief about what mattered most. The first thing that I had told Ruth I wanted in 1968 was to be a doctor, and after watching all my money and most of my friends disappear, I knew that being a doctor was my most powerful magic.

I was not sure exactly how to proceed following the dot-com crash or whether I wanted to continue in the role I had at Stanford as a clinical professor of neurosurgery. My interest in entrepreneurial activities was at its lowest then. I had in the past served as a consultant to hospitals that had difficulty providing neurosurgical coverage or were interested in developing neuroscience centers of excellence. I wanted there to be the best neurosurgical care possible, especially in areas where the majority of the population lived in poverty.

One day, out of the blue I was asked to advise a public hospital in southern Mississippi. As it was an hour from New Orleans, a city I loved and where I had gone to medical school, and it was a free trip, I said yes. The hospital was the primary provider of indigent care in the area, and as happens often many doctors didn't want to provide such care, as the reimbursement is very low. In addition, in this case, a private hospital run by a large hospital chain was incentivizing many of the specialists to practice at their institution, thus further exacerbating the situation. The problem was not only a lack of adequate neurosurgical coverage but a lack of coverage in the areas of neurology, orthopedics, and stroke care as well. I assessed the situation and explained to the hospital administration that there was a problem with the way they were making offers to potential doctors. They needed to explain that these doctors had the opportunity to be part of the development of a regional center of excellence. Not just to appeal to their egos but to that part of them that was present when they first became doctors . . . the ability to make a difference.

To create this regional center would require a large sum of

money. Following the presentation, the board unanimously voted to fund the vision to create a neuroscience regional referral center if I would agree to be the director of the program. It was an opportunity to lead an effort that would have a major impact in a place that really needed it. I surveyed colleagues and friends, none of whom could understand why I would voluntarily leave the weather of Northern California and the vibrant intellectual community of a major academic center. But after multiple visits to Mississippi, meeting wonderful people, and seeing a real need, I decided to make the move. In a fairly short period of time, I was able to recruit an extraordinary set of colleagues who enthusiastically engaged in the development of the center.

Many people in the United States don't appreciate that on almost all measures of quality or efficacy of healthcare, their country is in the last quadrant while having the most expensive care of all industrialized (first-world) countries and the least satisfied patients. What is also not appreciated is that every other industrialized country in the world offers universal healthcare to all of its citizens with better outcomes and much lower costs.

It has been shown that childhood poverty has a profound effect on one's health and ultimately future. Of course, I was well aware of this from firsthand experience, but when I moved to Mississippi this reality was again brought home to me. I remember being on call in the emergency room and seeing a child who had had a seizure and was now unresponsive, requiring a tube to be inserted into his windpipe to allow him to breathe. An emergency brain scan had been performed showing a large mass in his right temporal lobe compressing the normal structures of the brain and the brainstem. I spoke to the child's parents, who told me that he had been suffering for some time from an ear infection. Because they did not have insurance, the child was being seen by a nurse practitioner at a free clinic. He had repeatedly gone back because the antibiotics he had been given were not working, and he continued to complain of worsening ear pain and ultimately of a massive headache. They had no money to see a physician. The child had become confused and disoriented the day before, and they thought it was due to his fever. His parents finally took him to the emergency room after the seizure. To get there they had to call a neighbor to drive them since they didn't have a car.

I walked into the exam room and saw this beautiful child with a breathing tube on a ventilator. His frightened parents were at his bedside. I introduced myself and quickly examined the child, who had a widely dilated pupil on the right, and a slightly dilated pupil on the left. He was unresponsive, and his exam was consistent with pending brain death. I informed the parents that I had to act immediately to save the child's life and I asked them to leave the room. The scan had shown a mass that extended from the region of the right mastoid, that part of the skull that contains the ear canal, into the temporal lobe. With the child's history it was apparent that this child, whose ear infection should have easily been treated, had developed an infection of the mastoid bone that extended into the brain, resulting in a brain abscess. Such brain abscesses are rarely seen in this day and age. I quickly prepped and draped the child, clipped the hair over the temporal region, anesthetized the skin, incised the scalp, and drilled a burr hole overlying the area of the abscess. I then inserted a needle, and as I aspirated, pus filled the syringe. So much pus that I had to change the syringe three times.

I then took the child to the operating room, but it was too late. He was brain-dead. I left the operating room and walked into the waiting room. The parents stood. I could tell by how they looked that they were used to disappointment. I informed them that I had done everything I knew how to do to save their child's life and was not able to do so and that he was brain-dead—his body now only being kept alive by machines. After their tears and grief, they thanked me for trying, and my heart broke for all the times in their life people had not cared enough to try.

An ear infection or lack of health insurance should never cause a child's death.

Almost two years later, Hurricane Katrina struck. For many who had the ability to leave, it was an easy decision. Yet many more were stuck, stuck in a place of huge devastation, where recovery would take years if not decades. I struggled trying to decide whether I should leave or stay after the storm was over. I had come to assist the community, and I was enjoying caring for patients who truly needed help. We were building a resource for the community that would last into the future.

By this time, I had remarried a wonderful woman I had met shortly before giving my Accuray stock away. We had a young son, and my wife found it very difficult to live with my long hours and the daily reminders of devastation from Hurricane Katrina. Ultimately, we decided that she should move back to California with our child permanently, and I would remain in Mississippi but travel back and forth to California every six to eight weeks for a visit.

Many colleagues and friends couldn't understand why I just didn't leave permanently with my wife. The reality was that while that would have been easy, I couldn't face all those in the community, many of whom were now close friends and had believed in the vision that I had offered for that hospital becoming a regional referral center. For two more years I stayed and for several years after I remained deeply involved in this center, which became the center of excellence that I had envisioned so many years before. I finally left having built something that was, in fact, bigger than myself. After losing my wealth, I was committed to helping others, and this center, serving the needs of the poor, felt in a way like atonement for the years I had spent pursuing wealth and power.

As I was contemplating returning to California, I realized

I very much wanted to go back to Stanford. I also had been wondering what it was about Ruth's teachings that seemed so compelling and realized that at their core they were about opening the heart. Acting kindly and compassionately with intent. One of my fascinations was to understand how the brain and heart worked and interacted. Could compassion, kindness, and caring have signatures in the brain?

When I returned to Stanford on the neurosurgery faculty, I began meeting with colleagues in psychology and neuroscience to discuss what work was being done in this area. It turned out that there were a small number of researchers who were doing groundbreaking work on how being compassionate, altruistic, and kind affected the reward centers in the brain and positively affected their peripheral physiology. Compassion and kindness, it turned out, was good for your health. This research became my top priority, and I recommitted to the skills Ruth had taught me but developed them to better reflect the lessons I had learned. My notebook had been destroyed in Hurricane Katrina, when our house flooded, but I constantly replayed my conversations with Ruth in my head, hoping to gain new understanding, decades after the fact, about what Ruth had taught me. I immersed myself in the research that now was proving scientifically the benefit of all that Ruth had taught me. I wanted to study what it meant to open the heart and to understand why Ruth had emphasized this as being the most important. Just as I had made a list of my goals so many years before, I made another list of ten. A list of the ten things that open the heart.

I sat with it. I read it over and over, and then I suddenly saw it as a mnemonic, CDEFGHIJKL. It was a way to remember each aspect of what I had learned. The *alphabet of the heart*. While I continued the components of the meditation practice that I was taught in the back of the magic shop so many years before, I began a new practice each morning of reciting this new alphabet. After relaxing my body and calming my mind, I would recite this alphabet and set one quality from the list of ten as my intention for the day. I said them in my head over and over again. I found that it centered me, not only as a physician but also as a human being. It allowed me to start my day with a powerful intention.

THE ALPHABET OF THE HEART

C: Compassion is the recognition of the suffering of another with a desire to alleviate that suffering. Yet to be compassionate to another, you must be compassionate to yourself. Many people beat themselves up by being hypercritical, not allowing themselves to enjoy the same kindness that they would offer to others. And until one is truly kind to oneself, giving love and kindness to others is often impossible.

D: Dignity is something innate in every person. It deserves to be acknowledged and recognized. So often we make judgments about someone because of how they look, or talk, or behave. And many times such judgments are negative and wrong. We have to look at another person and think, "They are just like me. They want what I want—to be happy." When we look at others and see ourselves, we want to connect and help.

E: Equanimity is to have an evenness of temperament even during difficult times. Equanimity is for the good times and the bad times because even during good times there is a tendency to try to maintain or hold that feeling of elation. But trying to hold on to the good distracts us from being present in the moment just as trying to flee from the bad does. Grasping at that feeling of elation is not realistic, not possible, and only leads to disappointment. All such ups and downs are transient. Keeping an evenness of temperament allows for clarity of mind and intention.

F: Forgiveness is one of the greatest gifts one can give to another. It is also one of the greatest gifts we can give to ourselves. Many have used the analogy that holding anger or hostility against another you feel has wronged you is like drinking poison and hoping it kills the other person. It doesn't work. It poisons you. It poisons your interactions with others. It poisons your outlook on the world. Ultimately, it makes you the prisoner in a jail where you hold the key yet won't unlock the door. The reality is that each of us in our lives has wronged others. We are frail, fragile beings who at various times in our lives have not lived up to our ideal and have injured or hurt another.

G: Gratitude is the recognition of the blessing that your life is—even with all its pain and suffering. It takes little effort to see how so many in the world are suffering and in pain. People whose circumstances allow little hope of a better life. Too often, especially in Western society, we look at each other and feel jealous or envious. Simply taking a few moments to have gratitude has a huge effect on your mental attitude. . . . You suddenly recognize how blessed you are.

H: Humility is an attribute that for many is hard to practice. We have pride about who we are or what we have accomplished. We want to tell and show others how important we are. How much better we are than someone else. The reality is that such feelings are actually a statement of our own insecurity. We are searching for acknowledgment of worth outside of ourselves. Yet doing so separates us from others. It's like being put in solitary confinement, and it's a lonely place to be. It is only when we recognize that, like us, every person has positive and negative attributes, and only when we look at one another as equals, that we can truly connect. It is that connection of common humanity that frees us to open our heart and care unconditionally. To look at another as an equal.

I: Integrity requires intention. It requires defining those values that are most important to you. It means consistently practicing those values in regard to your interaction with others. Our values can easily disintegrate, and the disintegration can at first be imperceptible. If we compromise our integrity once, it becomes that much easier to do it again. Few start out with such intent. Be vigilant and diligent.

J: Justice is a recognition that within each of us there lives a desire to see that right be done. It is easier when we have resources and privilege to have justice. Yet, we need to guard justice for the weak and the vulnerable. It is our responsibility to seek justice for the vulnerable, to care for the weak, to give to the poor. That is what defines our society and our humanity and gives meaning to one's life.

K: Kindness is a concern for others and is often thought of as the active component of compassion. A desire to see others cared for with no desire for personal benefit or recognition. The extraordinary thing is that research is now finding that your act of kindness not only benefits those who receive your kindness but benefits you as well. The act of kindness ripples out and makes it more likely that your friends and those around you will be kinder. It is a social contagion that puts our society right. And ultimately kindness returns back to us, in the good feelings it generates and in how others treat us . . . with kindness.

L: Love when given freely changes everyone and everything. It is love that contains all virtues. It is love that heals all wounds. Ultimately, it is not our technology or our medicine but our love that heals. And it is love that holds our humanity.

. . .

THIS MNEMONIC connects me to my heart and allows it to open. It allows me to begin each day with intention and purpose. And throughout the day, when I am stressed or feel vulnerable, it centers me in the place I wish to be. It is the language of my intention. It is the language of the heart.

If Ruth were here, I think she might discover that I had finally learned to open my heart. And that has made all the difference.

THE HEART BEATS a hundred thousand times a day, sending two thousand gallons of blood through an intricate system of blood vessels that if stretched end to end would cover sixty thousand miles—more than twice the circumference of the earth. The ancient Egyptians believed that the heart—the *ib*—survived death, and in the afterlife, it passed judgment on the human who possessed it. The ancient Egyptian word for happiness is *awt-ib*, literally meaning "wideness of heart." The word for unhappiness was *ab-ib*, which meant "a truncated or alienated heart." In many cultures, both ancient and modern, the heart is seen as the seat of the soul and the secret place where the spirit dwells. When we read a story of a lost child, our heart can ache. When love ends, our heart can feel as if it might break and sometimes does. When we feel rejected, ashamed, or forgotten, our heart can feel tight and constricted, as if it were closing in on itself and getting smaller. But under pressure, whether from intense love or intense suffering, our heart can crack wide-open and never, ever be the same again. This is true not only in a metaphorical sense but in reality. In fact, there is actually a condition called *broken heart syndrome*.

It wasn't losing my money that cracked my heart wide-open—I found liberation in losing the wealth I had sought for so long—it was the pressure of keeping my heart closed for so long that finally caused it to break open. Ruth had said, "What you think you want is not always what's best." I had been chasing the wrong thing, and a heart ignored for too long will always make itself heard.

I also remembered my promise to Ruth: Someday I would teach this magic to others. I wasn't sure exactly how that would happen, but this was the focus of my visualization practice every night. Sometimes I saw myself in my white coat embracing a patient or a family member who was suffering, other times it was on a stage, and at other times I imagined myself talking to great philosophers and spiritual leaders. Even though I was, and am, an atheist, I thought often of my experience with Ruth and my experience after my car accident and found that I could have an open mind, be dogma-free, and still know that there is more to this life than I can explain. In many ways this was her gift as well. An acceptance that I don't need an absolute answer.

I feel that each of us is connected; when I look at another, I see myself. I see my weaknesses, my failings, and my fragility. I see the power of the human spirit, and the power of the universe. I know in my deepest being that it is love that is the glue that binds each of us. The Dalai Lama once said, "My religion is kindness," and that has become my religion as well.

I had always cared about others, and as a physician I care deeply for my patients. But the practice of opening one's heart with intention can cause pain. Pain so intense that at times it's almost unbearable. At times the pain didn't allow me always to be there or be as present as I wished. But when I truly open my heart as Ruth taught me, it actually changes how I respond to the pain. I did not need to run from it; I needed to be with it. And it was the being with it that allowed me to connect with myself and truly connect with others. My relationships with my patients have changed. I make more time to listen, and I try to open my heart to each one of them. I listen to their symptoms, and then I listen to their hearts—not with a stethoscope, but with my own heart.

THE STETHOSCOPE was invented because in 1816 a French physician was too embarrassed to put his ear up to a female patient's chest to listen to her heart and lungs (as was the norm at the time) and instead rolled twenty-four sheets of paper into a cone to create some distance between them. I think this distance between physician and patient has only grown larger over time. I learned that just by listening to my patients, just by giving them my time and attention and focus, they felt better. I let each of them tell his or her story, and then I acknowledged my patients' struggles, their accomplishments, and their suffering. And in many cases, this relieved their pain more than any medication I could offer and at times even more than my surgery. Even today, I tell my students and those residents I teach that while neurosurgery requires an immense amount of technology and sophisticated equipment, my greatest success as a neurosurgeon is the result of caring with an open heart and being present with my patients.

Another remarkable change was that everywhere I went, I saw people who were just like me. The clerk at the grocery store. The janitor who cleaned up the hospital late at night. The woman who stood at the traffic light holding a sign for money. The guy who was driving too fast in his Ferrari. And each of them had a backstory, just like me. Each of them was walking a path. Each of them struggled and suffered at times. From the person with the least to the person with the most, they were just like me.

I had begun to let go of the story that had defined my life. I had made an identity out of my poverty, and as long as I carried that identity with me, no matter how much wealth I accumulated, I would always be living in poverty. In my daily practice I opened my heart to my mother and father, and I found forgiveness for them. I opened my heart to the boy I used to be, and I found compassion. I opened my heart to all of the mistakes I had made and all the ways I had foolishly tried to prove my worth in the world, and I found humility. And in doing so, I knew that I wasn't the only one in the world to have been hungry. I wasn't the only one in the world who had ever been frightened. I wasn't the only one who had known loneliness or felt isolated and different. I opened up my heart and found that my heart had the ability to connect with every other heart it met.

It was exhausting and beautiful and strange.

All at the same time.

TWELVE

Manifesting Compassion

I have always enjoyed opera, although I can't say for sure why. Even without understanding a single word I often cry. Perhaps it is the raw emotion, the courageous display of passionate feeling that surpasses language. Opera isn't something you can intellectualize or explore with the mind—it can be felt only with the heart. Most surgeons play music in the operating room—it can calm and soothe the patient or focus and energize the surgical team. Studies have shown that when music is played to patients before surgery they exhibit less anxiety and require less pain medication and sedation. Like meditation techniques, music reduces heart rate, decreases stress, and lowers blood pressure. This calming effect happens not only for the patient but for the surgeon as well.

For me, if I play music during surgery the volume is low and the music is usually classical and calming during the critical phases of the surgery. As I am closing I might turn the volume up and play rock classics. But one type of music that I never play is opera. When I'm operating, I'm like a machine. My patients may want empathy and emotional connection before surgery, but during surgery, they want my skill, technical ability, and critical decision making. They don't want me crying over them on the operating table. They want me to care, but not if it gets in the way of saving their life.

June was one of my first patients in my new medical practice after I left my position as a neurosurgeon in the army, and June lived for opera. When she first swept into my office, she exuded vibrant energy and a warm spirit. She liked to wear high heels, and told me early on that she didn't care how great a doctor I was, she was never going to give up her two greatest passions, singing and pasta, even if I told her it would save her life.

June was a soprano in a traveling operatic troupe, and opera was both her vocation and the love of her life. We spent time during every appointment talking about her favorites—*Aida*, the Strauss operettas, and *Carmen*. Our appointments often lasted longer than usual because I enjoyed hearing her stories of singing around the country. She loved to make people feel.

"It sounds crazy, but I love it when my singing makes people cry—that's when I know I'm touching them. That's when I know I've connected."

June was having severe migraine headaches, and while the neurologist had been able to treat her headaches with medication, he couldn't fix the large aneurysm that was sitting adjacent to the left insula and that part of the brain associated with movement of the face region in her dominant hemisphere. It had been found as part of her medical workup for headache, and while it wasn't the cause of her headache, it had the potential not only to take from her what she valued so greatly but to kill her as well.

"Whatever's wrong with me," she said, "I don't want to do anything that will injure my voice or ability to sing—it's the most important thing I have."

I had to break the news to June.

The aneurysm, based on its size of over a centimeter in diameter, needed to be dealt with promptly, and I explained this to her over a number of appointments. I felt urgency but knew that June needed to have the delicate procedure explained to her slowly over and over again. I encouraged her to consult with other neurosurgeons, including colleagues who were much more experienced, even though I had done this surgery many times. Unfortunately, some neurosurgeons even with the most serious of conditions simply, and in a matter-of-fact fashion, describe the treatment and its associated risks without understanding that while routine for us, this treatment is often the most important event in the life of the patient and their family. The two other neurosurgeons she saw in second opinion were like this. She came back scared—with a sense that she was not a person but a diagnosis.

June needed this processing time, even more than most, and I tried to give her all the time that her condition allowed. Even back when I was a new physician, I knew that spending time with a patient was part of the art of medicine. Ultimately, we are dealing with real people who have real concerns and fears. Patients are not malfunctioning bits of machinery, and surgeons are not mechanics.

The more I talked to June, the more I saw her anxiety dissipate. She needed to tell her story, and she needed to know that I heard her story and knew her as a person. We developed a friendship. Ultimately, she told me I was the only one she trusted to do her surgery. While it is wonderful for a patient to have great confidence in your ability, it is different when a patient is a friend. The day before her surgery, she gave me a recording of her singing her favorite arias. That night I sat in my study with my eyes closed listening to her sing.

On the morning of June's surgery, I chose to play classic rock music from my childhood. She smiled at me warmly when she was wheeled in the operating room on a gurney and she heard the words of "All You Need Is Love" played through the speakers and they were the last words she heard as she drifted off to sleep. We transferred her from the gurney to the operating room table after she had been anesthetized, and I took the head clamp with its sharp pins and attached it to her head to secure it during the surgery. I could feel the pins penetrating her scalp and engaging the skull. I turned her head to the right and slightly extended her neck. I knew her appearance was very important to her and so I removed as little hair as possible. I reviewed the angiogram outlining the large bubble on the artery supplying a large part of the left side of her brain. It was an aneurysm arising at the bifurcation of the middle cerebral artery. I incised the scalp and turned the flap to reveal the skull. Normally the skull protects us, but in this case it was in the way. I used a craniotome to open her skull, which I then removed and placed carefully in a sterile towel. I could see her dura, that fibrous tissue that covers the brain, and knew that right beneath it was the aneurysm, keeping tune with the pulsating beat of her heart.

If it ruptured she could have a stroke and lose her voice or she could die.

I slowly opened the dura, and as I did, I could see the dome of the aneurysm poking out between the frontal and temporal lobes in the Sylvian fissure. I began the real work—bringing the microscope into position and using a micro-knife to dissect the delicate membrane from the brain surface, which allowed me to open the Sylvian fissure and gain access to the neck of the aneurysm, where the clip would be applied. I needed to separate it from her normal circulation. As I exposed the aneurysm, I saw that the wall was paper-thin. I could see, through the high-intensity light of the microscope, the blood swirling within the bulging and pulsating wall. It could have spontaneously ruptured at any moment. And part of the wall and neck were markedly attached to the surrounding brain, making them much harder to separate without rupture. Slowly, ever so slowly, I continued the dissection and was able to create a small path between the adherent scar tissue and the neck to allow me to place the clip. I didn't have even a millimeter of extra room. If I erred, it would rupture. My mistake could take away the one thing that meant the most to her—singing. I turned and reviewed the various clips, and placed one on the clip applier and turned toward that pulsating aneurysm that could kill her. I suddenly saw June's face in my mind and thought of her singing. I could hear her melodic voice. And then I thought of her paralyzed, unable to speak or sing. My hand holding the clip started shaking. Not a slight tremor but shaking. I couldn't proceed.

She was a friend. The woman who had told me her voice was the most important thing in the world. I had promised her nothing would happen. I had promised everything would be fine.

It's deadly for a surgeon to connect with a patient's humanity during surgery. It has to be a technical exercise. You have to objectify the person. If you think of what might happen to this fellow human being, you can't possibly do the surgery. It's too close to home. I felt scared. It had never happened before.

My hands were shaking so hard that I had to stop for a moment and sit down. I closed my eyes, and focused on my breath—inhaling then exhaling slowly—until I could create enough space in my thoughts that the fear had nothing to hang on to. There was a time to open my heart, and there was a time to rely on my skill and ability as a surgeon. My ability as an absolute technician. This was a procedure I had performed many times. One that I was extraordinarily good at. My fear left me, and I was back to that calm state of certainty about my intention. I could see in my mind the clip being placed and the aneurysm being obliterated. I turned back to June's open skull and focused the microscope back on the aneurysm, slowly guiding the clip into position in that tiny gap I had created, and once there, slowly closing its jaws. I then put a needle into the dome and drained the residual blood. It did not reexpand. The beast was truly dead and no longer a danger. June would sing again. I slowly closed the dura, replaced the bone flap, and closed the scalp. As I was putting the final head dressing on I realized the music was playing the same song that we had begun with. "Love is all you need, love is all you need."

June was extubated and taken to the recovery room. I sat down exhausted and closed my eyes for several minutes before I began writing orders. I thought of June and I thought of my hand shaking. I suddenly heard June's voice. "Where is Dr. Doty? I need to talk to him. I need to talk to him right now."

I walked over to her and took her hand. "Hi, June. How's it going?"

She looked deep in my eyes and saw what she needed to see. "Just fine, just fine. Thank you." Then she reached up to hug me and began crying as she realized she was going to be OK.

Driving away from the hospital a few hours later, I put in the CD that June had given me the day before. Just as the first few musical notes began, I accelerated onto the highway toward home.

June's voice suddenly filled the car with an aria from *Carmen*: *Habanera—Love is a rebellious bird*. I turned up the volume, rolled down my windows, and let the wind blow against my face. June had a gift. She could make people feel with her singing. She could touch people's hearts with her voice, and even through a recording, she could connect.

We all have that gift and ability to connect. Whether through music, or art, or poetry, or just through listening to another. There are a million little ways for our hearts to speak to each other, and this was June's way of reaching out to speak to mine.

The music made my heart ache. There was such beauty in her voice. I let my mind wander to what might have happened to June if the surgery hadn't gone well, and I could feel the tears well up in my eyes. I was grateful she would be able to continue sharing her gift with the world, and that gratitude brought even more tears. I couldn't sing opera, but I could still feel how much it meant to her. At that moment I wanted to be home. I wanted to hug those I loved. And I was thankful. Thankful that I was able to help June. Thankful that I was a doctor.

IT CAN HURT to go through life with your heart open but not as much as it does to go through life with your heart closed. I was still struggling with how to reconcile the part of me that had to be a detached neurosurgeon with the part of me committed to connecting with others.

I found myself thinking often of Ruth and wishing I could ask her as an adult the same thing I had asked her as a child: Why? What made Ruth reach out to me, when so many don't reach out? Ruth wasn't wealthy, and she wasn't without her own life problems, but her heart was open, and she saw someone who was in need and did something about it. It made me wonder, how is it that those who have so much can do so little to help those that are struggling? And how is it that some, who have nothing by way of material things, will still offer everything they have to someone less fortunate? Why will some people, like Ruth, go out of their way to help, and why do others turn their backs on someone who is suffering?

These weren't just idle philosophical reflections. I began

devoting myself to rigorous scientific research and collaborating with others who were exploring similar areas. I had explored the mysteries of the brain, and it was time to devote as much academic rigor and hard science to exploring the secrets of the heart.

What I have learned since is that compassion is an instinct, perhaps our most innate. Recent research shows that even an animal can go through tremendous effort and cost to help out another of its species—or even of another species—who is suffering. Monkeys care for each other when they're injured, baby owls feed their less fortunate nest mates with bits of their own food, a dolphin has even helped save a beached humpback whale. We humans are even more instinctually compassionate; our brains are wired with a desire to help each other. We see this desire to help in children as young as toddlers.

There is a part of our brain called the central or periaqueductal grey matter, and its connections to the orbitofrontal cortex are responsible in great part for nurturing behavior. When we see others in pain or suffering, this part of the brain activates, meaning we are wired to nurture and help others when they are in need. Similarly, when we give to others, it lights up the pleasure and reward centers in the brain, even more so than when someone gives to us. And when we see someone acting kindly or being helpful, this in turn causes us to act more compassionately.

Many misinterpret Darwin by implying that survival of the

fittest means the survival of the strongest and most ruthless, when in fact it is survival of the kindest and most cooperative that ensures the survival of a species in the long-term. We evolved to cooperate, to nurture and raise our dependent young, and to thrive together and for the benefit of all.

I cried over June that day, just as I have cried over other patients since, although I have never again had a surgery interrupted by such emotion. There's no shame in caring or feeling someone else's pain. It is beautiful and, I think, why we are all here in this life together.

WHILE WRITING THIS BOOK, I found out that Ruth had died in 1979 from breast cancer, so while I'll never know for sure, I do believe that Ruth would have been proud of my quest to open my own heart and the hearts of others. And I think she would have understood my desire to prove scientifically that which she knew intuitively. When our brains and our hearts are working in collaboration—we are happier, we are healthier, and we automatically express love, kindness, and care for one another. I knew this intuitively, but I needed to validate it scientifically. This was the motivation to begin researching compassion and altruism. I wanted to understand the evolution of not only why we evolved such behavior but also how it affects the brain and ultimately our health. Clearly, there was preliminary evidence that showed significant positive effects. My goal was to join a small group of researchers who had already been working in this area. On a personal level I already knew the effect, but wondered if we could create ways to improve people's lives through this knowledge. Could I contribute?

I had already begun some preliminary investigations with colleagues in neuroscience and psychology. The results were encouraging. We had even begun meeting every few weeks to discuss the latest research as well as potential research projects. We called this informal initiative Project Compassion. Initially, I was funding this research myself. During one of our meetings the Dalai Lama's name had come up, as one of the leading centers doing this work had been encouraged by him to research the effects of meditation and compassion on the brain. A few days later while walking through the Stanford campus, a vision of the Dalai Lama just popped into my head. I thought, wouldn't it be great to have him come to Stanford, meet with me and my colleagues, and talk about compassion. It's interesting, because I wasn't a Buddhist, nor did I know much about the Dalai Lama other than that he had visited Stanford in 2005 and discussed addiction, craving, and suffering. Yet, I couldn't get the idea of him visiting again out of my head. I found out that the visit in 2005 had been, in part, motivated by the dean of the medical school's wife, who was an admirer of the Dalai Lama. She told me that one of the faculty members in the Stanford Tibetan Studies Initiative had been responsible for making the appropriate introductions. I contacted him, and he was very encouraging. He referred me to the Dalai Lama's English translator, Thupten Jinpa, a former monk who had been working with His Holiness for almost a quarter of a century at that time. He and I spoke on the phone, and he arranged for a meeting with the Dalai Lama during his visit to Seattle in 2008.

And just like that, I had manifested the Dalai Lama.

Several Stanford representatives accompanied me on my trip to Seattle—a representative from the school of medicine, the dean for Religious Life, the director of the Stanford Neurosciences Institute, the Tibetan studies professor who had arranged the first connection, and a potential benefactor. It was quite an entourage, and one I had not quite planned on when I had the idea to have the Dalai Lama come speak.

We met in his hotel room and introductions were made, after which I explained to His Holiness my interest in compassion and my background as a physician and neurosurgeon, the preliminary research that we had recently begun on compassion, and my desire to have him speak at Stanford. He asked several insightful questions about the research and the science of compassion. After I finished answering he looked at me and smiled. He said, "Yes, of course I will come."

It is quite extraordinary to be in the presence of the Dalai Lama. There is this absolute and unconditional love he exudes that feels just like taking a deep breath after holding your breath for a long time. You don't have to be anyone other than who you are, and you are met with total acceptance. It's a profound feeling, and there are no words that can adequately explain it. A monk soon brought out a large paper ledger to find space in the calendar to schedule the visit. A date was agreed upon. Suddenly the Dalai Lama began an intense and animated discussion in Tibetan with his translator. This went on for quite a while, and the Stanford entourage all sat silently. Had I done something to upset him? Had I inadvertently pissed off the Dalai Lama? What were they saying?

I began to sweat and feel anxious.

The conversation abruptly came to an end, and his translator, Jinpa, turned to me and said, "Jim, His Holiness is so impressed by your intent and this endeavor that you have begun that he wishes to make a personal contribution to your work."

He told me the amount, and I was dumbfounded. This was extraordinary and unprecedented. The Dalai Lama does have discretionary funds from the sale of his books that typically he gives to Tibetan causes or initiatives. He had given smaller amounts in the past to various causes, but this donation turned out to be the largest sum he had ever given to a non-Tibetan cause. The meeting ended with all of us feeling as if we were floating on a cloud. Not only had His Holiness agreed to come speak at Stanford, but he was now our benefactor. Amazing. Afterward one of the individuals at the meeting told me that based on how His Holiness had responded to me that he felt compelled to make a donation to my work. A week later an engineer from Google whom I had met and was interested in my work called to say he had heard about the meeting and was so impressed with the donation by His Holiness that he also wanted to contribute. Ultimately, all three made incredible monetary contributions. What had begun as an informal project now became formalized by the dean of the medical school, with support from the director of the Neurosciences Institute and the chairman of my department, as the Center for Compassion and Altruism Research and Education (CCARE). And just as extraordinarily, Jinpa, who in addition to being a former monk had a PhD from Cambridge, ended up becoming a close friend and spending a week every month for the next three years helping me create what is today CCARE. At the same time, with colleagues from psychology, he helped develop a training program for cultivating compassion, which has now been taught to thousands and which we continue to research in regard to its effect. We have also trained instructors who have brought the power of this training to many parts of the world and who no doubt will bring it to many more over the years.

Since its founding, CCARE has been recognized as a pioneer and leader in the field of compassion and altruism research and has promoted the profound effect such behaviors can have on the lives of individuals, in education, in business, in healthcare, in social justice, and in civic government. We hope it will serve as a beacon of light, demonstrating the power of an individual to affect the lives of others and further showing empirically the value of these behaviors in terms of health, wellness, and longevity.

I had a personal experience with the power of an individual to affect the life of another. It is my hope that CCARE will inspire others to know the same kind of power. CCARE is one way of doing what Ruth asked me to do—teach her magic to others. Guiding other physicians is another.

THIRTEEN

The Face of God

More than twenty-five thousand years ago, Hippocrates, considered in Western culture to be the "Father of Medicine," required each of his students to take an oath swearing to abide by the highest ethical standards as they practiced the profession of medicine. Many people remember the Latin phrase *Primum non nocere*, "First do no harm," as a core tenet of medicine, believing that Hippocrates was the first to have uttered the words—but they would be wrong. The phrase is believed to have originated with Thomas Sydenham, a seventeenth-century English physician, who wrote a textbook of medicine used for two hundred years that resulted in his being called the "English Hippocrates."

Over the last two decades in the United States and in many

parts of the world, the tradition of medical students taking the oath of Hippocrates immediately before the start of classes has been formalized into what is known as the "White Coat Ceremony," when the students are given white coats and recite the oath, after which an individual who epitomizes the highest ideals of medicine gives an inspirational speech welcoming the students to the profession.

Thirty years after I graduated from Tulane Medical School in New Orleans, the dean of the school that accepted me without a degree and with the lowest GPA of anyone attending, called to ask me to be that speaker. I cannot tell you the emotions that went through me as I heard the words. Me, Jim Doty, the failing undergraduate who had been told that applying to medical school was "a waste of everyone's time," asked to be the speaker at the White Coat Ceremony at my alma mater and being held up as a role model to a whole class of aspiring student physicians?

I am frequently amazed at where life has taken me.

It's easy to connect the dots of a life in retrospect, but much harder to trust the dots will connect together and form a beautiful picture when you're in the messiness of living a life. I could never have predicted either the successes or the failures in my life, but all of them have made me a better husband, a better father, a better doctor, and a better person.

I have taken my role as a healer with great seriousness. The lessons that Ruth taught me allowed me to open my heart and temper that seriousness with kindness and compassion. Not only did her magic allow me to believe that I could attend college and medical school but it gave me the tools to complete neurosurgical training, one of the most difficult and arduous residencies in medicine, and to become a professor at one of the most prestigious medical schools in the country.

The magic also gave me the courage to take risks and feel secure that, regardless of the outcome, I would be OK. The risk of taking over a failing medical device company and putting everything on the line because of a belief in the importance of the technology in saving lives. The risk of giving away what it was that I thought I wanted most—money—the very thing I thought would make me happy and give me control in life. Her magic made me realize that it was OK to be me, money or not, and that in reality none of us has control. I had been chasing a chimera, and letting that go gave me the most valuable gifts of all: clarity, purpose, and freedom.

Like the Dalai Lama, my religion is kindness. It is a religion that doesn't require a god who sits in judgment or lengthy dogmatic texts. It is also a religion that doesn't allow for anyone to feel superior to another and requires us to accept that we are all equal. This religion inspired me to research how compassion and kindness are critical to one's mental and physical health and longevity.

As I prepared for the speech, I thought of all these things and more. What could I give these students who were just beginning the arduous journey of becoming a physician? What could I give them that they could carry with them over the course of their careers? I thought of Ruth and the lessons she taught me that are with me every day. I thought of the mnemonic that had proved so powerful to me and that I recited every morning after awakening and often several times throughout the day. I thought of the patients I had met who taught me how to care and how to love. And I thought of death and how we have so little time in this world.

I had learned to relax my body, quiet my mind, open my heart, and visualize what I wanted to manifest. I learned that what I wanted to manifest most was a world where people not only did no harm to one another but reached out to help one another. I had learned to use the compass of my heart to guide my way and to trust that wherever I ended up, that was exactly where I needed to be. I learned that we all fundamentally have the same brains and the same hearts and the same ability to change them, transform them, and use them for the benefit of all. I learned not to define people by where they are born, what they do, or how much they have. And I learned not to define myself by this either. I once thought there was something wrong with me because of the nature of my circumstances. I believed I had no worth if I had no money. I realized that I was not responsible for the circumstances of my birth and that to be defined by them was wrong. Everyone has worth, has value, and deserves to be treated with dignity and respect. Everyone deserves love. And everyone deserves a chance, and then a second chance.

Each of us has a story, and in each story there are parts that are painful and sad. We can choose, at any moment, to see the people right in front of us for who they are and who they can be. Ruth saw a scared and lonely boy, but she also saw within me a heart that had been hurt. Each of us has wounds. And each of us has the ability to heal. She helped me heal. And you can do the same. Giving love is always possible. Every smile at a stranger can be a gift. Every moment of nonjudgment of another human being is a gift. Every moment of forgiveness, for yourself or for someone else, is a gift. Every act of compassion, every intention to serve, is a gift to this world and a gift to yourself.

We are at the beginning of an age of compassion. People are yearning for an understanding of their place in the world and a way to be content and happy, and they are looking for a method of transformation. Ruth taught me a method that worked for me and maybe it was her insight and skill that allowed it to manifest as it did. Others have found their own methods to quiet their minds and open their hearts. Right now it's a ripple in human consciousness fueled by compassion, but it's a ripple that has the potential to become a tsunami.

We are on a journey of connection. It is the journey of opening our heart to our fellow beings on this earth and recognizing that they are our sisters and brothers. Recognizing that one act of compassion leads to another act of compassion, and so on across the globe. In the end, how well we love each other and how well we take care of each other will be what determines the survival of our planet and our species. His Holiness the Dalai Lama says, "Love and compassion are necessities; without them humanity cannot survive." I realized not only in medicine but in life this was true. How was I going to share these values with this group of young students who were about to embark on a career of service?

I walked up the steps to the stage at the auditorium at Tulane and looked out at the twelve hundred students, faculty, and their families. I scanned the expectant faces of the students. I recalled sitting in the auditorium attending my own version of the White Coat Ceremony so many years before, but sadly I could neither recall the speaker nor what was said. In fact, my only recollection was receiving a white coat and taking the oath.

I began to speak as a great wave of emotion washed over me. I shared with the audience my journey and told them of the doctor who inspired me in the fourth grade and of the woman who believed in me, Ruth. Each of them listening, I said, had the power to change the life of others for the better, not only the lives of their patients but also the lives of all those around them. Sometimes it takes only a smile or a kind word. I told them that while medicine had changed, it was still a noble profession. Then I told them about the alphabet of the heart and I went through each letter and its meaning. As I finished with *L* and the word *love*, my voice cracked, and I could feel tears welling up in my eyes.

"There is no perfect life we are born into, and there is no escaping the awful reality of suffering. There is also no escaping the beautiful synchronicity of the heart." I paused for a moment as I prepared to end my talk. I saw a young man in the audience and saw myself all those years ago.

"Today you have sealed your path with an oath. This path will take you to life's deepest and darkest valleys where you will see how trauma and disease destroy lives, and sadly you will see what one human is capable of inflicting upon another and even more sadly what one human is capable of inflicting on himself. But it will also take you to life's highest peaks where you will see the meek demonstrate strength you thought not possible, cures for which you can find no explanation, and the power of compassion and kindness to cure human ills. And by doing so you will see the very face of God."

I realized I had been so focused on these last words that I wasn't paying close attention to the audience. As I finished, I saw that many were crying. I looked around at my colleagues onstage and they too were crying. And I realized that I had tears running down my cheeks. Suddenly the entire audience stood and applauded. They were not applauding just me or my journey but our collective journey toward greater compassion and ever-greater humanity.

So many people were waiting at the side of the stage, thanking me, crying, and telling me how my talk had opened their hearts.

I thought of my life and of Ruth. I again realized the power of her words and the power of her magic. It is a power that lives in each of us that is just waiting to be released. It is the gift we can give to one another.

I walked out of the auditorium and felt the warmth of the sun on my face. I paused and closed my eyes and allowed myself to just be.

It was OK.

I was OK.

I began my quest to discover the mysteries of the brain and the secrets of the heart in a magic shop, but the truth is, we don't need to walk into a magic shop to discover them. We need only to look into our own minds and into our own hearts.

Now it is up to you to make your own magic. And to teach others. The brain and the heart, working together, can make the most extraordinary magic there is. It has nothing to do with illusions or sleights of hand.

This magic is real.

And just as it was the greatest magic Ruth could offer me, it is also the greatest magic I can offer you.

ACKNOWLEDGMENTS

In my position as the founder and director of the Center for Compassion and Altruism Research and Education (CCARE) at Stanford University School of Medicine, I have shared many times the story of my childhood and what motivated me to dedicate a great portion of my time and energy to researching compassion and its power to change lives. The stories I shared seemed to resonate deeply with many people, and often I was asked when I would write a book. For many reasons, I had avoided such entreaties in part because it required a commitment of time and effort in the face of an already busy schedule, and probably more so because I knew from telling the stories that often they took me back to periods in my life that were difficult and painful.

My feelings changed when, while attending the eightieth birthday of Desmond Tutu in Cape Town, I had the privilege of meeting Doug Abrams of Idea Architects. At the time, I was not aware that he was Archbishop Tutu's literary agent. Unbeknownst to me then, Doug had attended many CCARE events. He shared with me how inspiring he found my stories, and he thought that a book had the capacity to inspire many and shared with me how inspiring they had been to his father. In fact, he told me, the reality is that while his goal as a literary agent was to bring inspiring stories to the world, the bigger motivator was to bring this narrative to his father in the form of a book. How could I say no?

ACKNOWLEDGMENTS

Like so many things in life, they are not done alone or by one individual. And this is the case in this instance. Not only was Doug critical in assisting me in creating a proposal, more important, through his contacts and the respect he garners in the publishing world, he was able to partner me with the extraordinary Caroline Sutton at Avery, an imprint of Penguin Random House. Her support, encouragement, and guidance really allowed my story to come to life in the form of a book.

Once the contract was signed, I suddenly realized the burden I had accepted and the associated deadline for completion. Fortunately, Idea Architects came to the rescue by partnering this effort with their editorial director, Lara Love. At every stage, I could not have asked for a more helpful, diligent, and thoughtful person to guide me through the process of the writing and editing. Also her ability to turn a phrase, her talent for finding the critical details that bring a story to life, and her gentle nudging for me to go to places that often were uncomfortable and painful were critical to any success this book might achieve. For almost two years, Lara and I met twice weekly before sunrise, and it is through this period that she also became a close friend, and it is this friendship for which I am most grateful.

I would also like to thank my extraordinary wife and life partner, Masha, whose support I try not to take for granted. Being married to a neurosurgeon means many missed events and often leaving in the middle of the night and returning home exhausted. In the face of this, my wife has supported my endeavors in promoting the power of compassion to change lives. For this I am forever grateful.

I would like to acknowledge the many others who throughout my life have helped me along my journey and who have so often shown me the path.